Annual Reports in
COMPUTATIONAL CHEMISTRY

VOLUME **4**

Annual Reports in
COMPUTATIONAL CHEMISTRY

VOLUME **4**

Edited by

RALPH A. WHEELER
Department of Chemistry & Biochemistry
University of Oklahoma
620 Parrington Oval, Room 208
Norman, OK 73019
USA

DAVID C. SPELLMEYER
Nodality, Inc.
7000 Shoreline Court, Suite 250
South San Francisco, CA 94080
USA

Sponsored by the Division of Computers in Chemistry
of the American Chemical Society

ELSEVIER   Amsterdam • Boston • Heidelberg • London • New York • Oxford
Paris • San Diego • San Francisco • Singapore • Sydney • Tokyo

Elsevier
Linacre House, Jordan Hill, Oxford OX2 8DP, UK
Radarweg 29, PO Box 211, 1000 AE Amsterdam, The Netherlands

First edition 2008

British Library Cataloguing in Publication Data
A catalogue record for this book is available from the British Library

Library of Congress Cataloging-in-Publication Data
A catalog record for this book is available from the Library of Congress

ISBN: 978-0-444-53250-3
ISSN: 1574-1400

For information on all Elsevier publications
visit our website at elsevierdirect.com

Printed and bound in USA

08 09 10 11 12   10 9 8 7 6 5 4 3 2 1

CONTENTS

CONTRIBUTORS

Evan E. Bolton
National Center for Biotechnology Information, National Library of Medicine, National Institutes of Health, Department of Health and Human Services, 8600 Rockville Pike, Bethesda, MD 20894, USA

Stephen H. Bryant
National Center for Biotechnology Information, National Library of Medicine, National Institutes of Health, Department of Health and Human Services, 8600 Rockville Pike, Bethesda, MD 20894, USA

David A. Case
Department of Molecular Biology, The Scripps Research Institute, La Jolla, CA 92037, USA

Alan C. Cheng
Amgen Inc., One Kendall Square, Bldg 1000, Cambridge, MA 02139, USA

Michael Feig
Department of Biochemistry and Molecular Biology, Michigan State University, East Lansing, MI 48824, USA

Eric Franzosa
Bioinformatics Program, Boston University, 24 Cummington Street, Boston, MA 02215, USA

Rigoberto Hernandez
Center for Computational Molecular Science and Technology, School of Chemistry and Biochemistry, Georgia Institute of Technology, Atlanta, GA 30332-0400, USA

Robert Ezra Langlois
Bioinformatics Program, Department of Bioengineering, University of Illinois-Chicago, 851 S Morgan, Room 218, Chicago, IL 60607, USA

Andrew R. Leach
Medicines Research Centre, GlaxoSmithKline Research & Development, Gunnels Wood Road, Stevenage, Hertfordshire SG1 2NY, UK

Alessio Lodola
Dipartimento Farmaceutico, Università degli Studi di Parma, viale G.P. Usberti 27/A Campus Universitario, I-43100 Parma, Italy

Hui Lu
Bioinformatics Program, Department of Bioengineering, University of Illinois-Chicago, 851 S Morgan, Room 218, Chicago, IL 60607, USA

Adrian J. Mulholland
Centre for Computational Chemistry, School of Chemistry, University of Bristol, Bristol, BS8 1TS, UK

Jens Erik Nielsen
School of Biomolecular and Biomedical Science, Centre for Synthesis and Chemical Biology, UCD Conway Institute, University College Dublin, Belfield, Dublin 4, Ireland

Alexey Onufriev
Department of Computer Science and Physics, 2050 Torgersen Hall, Virginia Tech, Blacksburg, VA 24061, USA

Alexander V. Popov
Center for Computational Molecular Science and Technology, School of Chemistry and Biochemistry, Georgia Institute of Technology, Atlanta, GA 30332-0400, USA

Sanbo Qin
Department of Physics and Institute of Molecular Biophysics, Florida State University, Tallahassee, Florida 32306, USA

Yanping Qin
Center for Computational Molecular Science and Technology, School of Chemistry and Biochemistry, Georgia Institute of Technology, Atlanta, GA 30332-0400, USA

Maryam Sayadi
Department of Chemistry, Michigan State University, East Lansing, MI 48824, USA

Stefan Senger
Medicines Research Centre, GlaxoSmithKline Research & Development, Gunnels Wood Road, Stevenage, Hertfordshire SG1 2NY, UK

Seiichiro Tanizaki
Department of Chemistry and Biochemistry, University of Texas at Arlington, Arlington, TX 76019, USA

Paul A. Thiessen
National Center for Biotechnology Information, National Library of Medicine, National Institutes of Health, Department of Health and Human Services, 8600 Rockville Pike, Bethesda, MD 20894, USA

Harianto Tjong
Department of Physics and Institute of Molecular Biophysics, Florida State University, Tallahassee, Florida 32306, USA

Yanli Wang
National Center for Biotechnology Information, National Library of Medicine, National Institutes of Health, Department of Health and Human Services, 8600 Rockville Pike, Bethesda, MD 20894, USA

Christopher J. Woods
Centre for Computational Chemistry, School of Chemistry, University of Bristol, Bristol, BS8 1TS, UK

Vance Wong
Department of Molecular Biology, The Scripps Research Institute, La Jolla, CA 92037, USA

Yu Xia
Bioinformatics Program, Boston University; Department of Chemistry, Boston University, 24 Cummington Street, Boston, MA 02215, USA

Huan-Xiang Zhou
Department of Physics and Institute of Molecular Biophysics, Florida State University, Tallahassee, Florida 32306, USA

Annual Reports in Computational Chemistry (ARCC) focuses on providing timely reviews of topics important to researchers in the field of computational chemistry. The *ARCC* is published and distributed by Elsevier and is sponsored by the Division of Computers in Chemistry (COMP) of the American Chemical Society. Members in good standing of the COMP Division receive a copy of the *ARCC* as part of their membership benefits. We are very pleased that all three previous volumes have received a very positive response from our readers. The COMP Executive Committee expects to deliver future volumes of the *ARCC* that build on the solid contributors of our first four volumes. To ensure that you receive future installments of this series, please join the Division as described on the COMP web-site at http://membership.acs.org/c/Comp/index.html.

In Volume 4, our Section Editors have assembled 12 contributions in five sections. Topics covered include a new section on Bioinformatics (Wei Wang), as well as continuing sections on Simulation Methodologies (Carlos Simmerling), Biological Modeling (Nathan Baker), Physical Modeling (Jeffry Madura), and Emerging Technologies (Wendy Cornell). We anticipate that two sections included in prior volumes, Quantum Chemistry and Chemical Education will reappear in the next volume, Volume 5. With Volume 4, we extend the practice of cumulative indexing of both the current and past editions in order to provide easy identification of past reports.

As was the case with our previous volumes, the *Annual Reports in Computational Chemistry* has been assembled entirely by volunteers in order to produce a high-quality scientific publication at the lowest cost possible. The Editors extend our gratitude to the many people who have given their time to make this edition of the *Annual Reports in Computational Chemistry* possible. The authors of each of this year's contributions and the Section Editors have graciously dedicated significant amounts of their time to make this Volume successful. This year's edition could not have been assembled without the help of Deirdre Clark, of Elsevier. Thank you one and all for your hard work, your time, and your contributions.

We hope that you will find this edition to be interesting and valuable. We are actively planning the fifth volume and are soliciting input from our readers about future topics. Please contact either of us with your suggestions and/or to volunteer to be a contributor.

Sincerely,

Ralph A. Wheeler and David Spellmeyer, Editors

Section 1
Bioinformatics

Section Editor: Wei Wang

Department of Chemistry and Biochemistry
University of California at San Diego
La Jolla, CA 92093
USA

CHAPTER 1

Structural Perspectives on Protein Evolution

Eric Franzosa* and **Yu Xia***,****

Contents

1. INTRODUCTION

Nothing in biology makes sense except in the light of evolution [1]. These words, written by famed evolutionary biologist Theodosius Dobzhansky, have a special place

* Bioinformatics Program, Boston University, 24 Cummington Street, Boston, MA 02215, USA
** Department of Chemistry, Boston University, 24 Cummington Street, Boston, MA 02215, USA. E-mail: yuxia@bu.edu

Annual Reports in Computational Chemistry, Vol. 4
ISSN 1574-1400, DOI: 10.1016/S1574-1400(08)00001-7

in the life sciences. They serve not only as a reminder of the central role evolution plays in the processes of life, but also as a paradigm under which research in biology should be conducted. When we think about evolutionary phenomena, it is important to remember that they—like all natural phenomena—can be reduced to events at the molecular scale. Evolutionary change is fueled by mutations: changes in the molecular structure of the genetic material. When these mutations are expressed, they result in changes to biomolecular structure and energetics which may in turn alter the abundance or interactions of proteins. Novel organismal traits stemming forth from these molecular scale changes are then judged by natural selection. Considering this perspective, it seems that nothing in evolution makes sense except in the light of biomolecular structure and energetics. It should therefore not come as a surprise that the relationship between evolution and physical phenomena like these has been an area of interest for some time. Here we review some of the major theories, models, and empirical evidence relevant to the relationship between protein structure and evolution at various scales.

2. DETERMINANTS OF EVOLUTIONARY RATE

Protein evolutionary rates are known to vary widely. In the genome of the model organism *Saccharomyces cerevisiae* (baker's yeast), evolutionary rates among the roughly 6,000 genes are spread out over three orders of magnitude [2]. Since the advent of the molecular biology age scientists have been interested in the way that homologous genes and proteins accumulate changes. It has been observed that the sequences and structures of some proteins are highly conserved, even when comparisons are made between distantly diverged species (for example, the histone proteins that package DNA, or the ribosomal proteins responsible for protein translation) [3]. Other proteins evolve rapidly, either due to relaxed constraint or positive selection for novel features (for example, proteins involved in immune systems) [4]. While theories explaining these differences originated alongside their observation, extracting general determinants of evolutionary rate variation has only become possible with the advent of the bioinformatics age. Statistical and machine learning techniques, when applied to massive genomic and phenotypic datasets (such as protein structures, interaction networks, and expression profiles), have been able to isolate some of the forces driving evolution at the molecular level (for general reviews of evolutionary determinants, see [2,5,6]).

Features that are directly connected to protein structure have been shown to explain roughly 10% of the variation in evolutionary rate [7]. This result seems initially surprising, given that structure mediates all aspects of a protein's existence. In contrast, expression—the frequency and scale at which a protein is manufactured—may explain up to 40% of evolutionary rate variation [6]. Expression and evolutionary rate vary inversely, with highly expressed proteins tending to evolve at very slow rates. A protein's dispensability (effect on cell growth when absent) and the number of interactions in which it participates explain additional components of evolutionary rate variation; random noise may also be a large contributor [8]. It is worth noting that protein function, historically considered to be a major target of selection, does not seem to be a good general predictor of evolutionary rate [6]. While much progress has been made in the identification and ranking of evolutionary determinants, some disputes in this area still remain. We believe that considering structure is particularly important because of its role as both a determinant of evolution in its own right, and a medium through which other determinants act.

As an example, we will consider the role that structure plays in the apparent dominance of expression in the determination of evolutionary rate. Although several hypotheses have been proposed to explain the significance of expression [9,10], genomic evidence seems to best support the following conclusion:

Errors made during protein translation can result in misfolded proteins, which represent a burden to the cell. Mutations that make a protein more susceptible to error-induced misfolding will result in a loss of fitness. If the mutation occurs in a highly expressed protein, then translational errors (and misfolding events) will be more common, resulting in a larger fitness loss. Hence, protein expression will scale directly with selective constraint, and inversely with evolutionary rate [11].

Are errors in translation really so common that they can have a profound influence on the evolutionary trajectory of a protein? Although the machinery of translation operates with 99.95% accuracy (measured as correctly inserted amino acids), even a small potential for error becomes rapidly compounded given the enormous work load it must handle [12]. If we assume that an average protein is composed of 400 amino acids, roughly 20% of these proteins will contain at least one translational error. Robustness against error-induced misfolding (i.e., structural robustness) would presumably be beneficial for any protein, but more so for the highly expressed among them. Thus, structure, a characteristic of all proteins, plays an even more critical role in their evolution than is apparent at face value. A theoretical treatment of the sequence-structure relationship sheds light on the role of structure in this and other evolutionary phenomena.

3. THEORETICAL ADVANCES

3.1 Key concepts

The essence of most theoretical ideas governing protein structure and evolution begin with the following relationship:

$$\textbf{Genotype} \quad \rightarrow \quad \textbf{Phenotype}$$
$$\text{(sequence)} \qquad \qquad \text{(structure)}$$

Genotype yields phenotype. This is a general biological idea. In the case of proteins, it can refer to the DNA sequences (genotypes) that encode amino acid chains that fold to produce three-dimensional proteins (phenotypes). Alternatively we can bypass the genetic component of the picture and think of the translated sequence of amino acids as a genotype, with the notion of a phenotype remaining the same. The relation above is simple, but profoundly important. In one sense it can be thought of as a statement of the central dogma of molecular biology (the elucidation of which is among the greatest scientific achievements of the 20th century) [13]. In another sense this relation is a statement of the protein folding problem, one of the largest challenges facing researchers in computational biology today [14].

The space of protein phenotypes observed in nature is surprisingly small. Current estimates place the number of stable folds in the neighborhood of 1000 to 10,000 [15,16]. We can imagine many other possible folds in the configuration space of an amino acid chain, but these have either (a) yet to occur in evolution or (b) been thermodynamically or selectively disfavored. The complete space of possible genotypes is assumed to be very large. Constraining ourselves to the size of an average protein (400 amino acids), there are 20^{400} ($\approx 2.6 \times 10^{520}$) possible protein sequences. Obviously evolution has sampled only a small fraction of these sequences, and an even smaller fraction persists on the planet today. Nevertheless, the mapping of genotypes to phenotypes remains many-to-one, with sets of genes and amino acid sequences producing largely identical protein structures [17]. We make the assumption that, under a given set of conditions, a single sequence maps to exactly one structure (governed by the minimization of free energy). This is a reasonable assumption for natural protein sequences, which tend to have a marked free energy minimum [18].

These observations are extremely important in light of the neutral theory of evolution. Simply put, this theory states that the majority of accepted changes that occur at the genotype level do not have a pronounced effect on the phenotype [19]. Silent substitutions in DNA are one example of this phenomenon. DNA codons **GGA** and **GGG** both encode the amino acid glycine, and hence a **GG<u>A</u>** → **GG<u>G</u>** mutation would produce a genotype change, but not a phenotype change. Note how this idea fits naturally with the observation of the many-to-one mapping of protein sequences to structures. Since a given structure may be generated by multiple sequences, mutations that interconvert those sequences do not have phenotypic consequences, and are therefore selectively neutral. Some caution is warranted here, as no mutation is likely to be neutral across all environments [20]. We can imagine that sequences which produce identical structures under one temperature regime might produce two different structures under another. Even the canonical silent DNA polymorphisms can evolve non-neutrally in situations where one synonymous codon is preferred over another for purposes of transcriptional or translational efficiency [21]. In this review, we frequently employ the approximation that for most mutations, protein structure directly dictates protein function, i.e. mutations that preserve a protein's structure will also preserve its function. This is not always the case, as certain mutations which conserve structure may have significant functional consequences (for example, if they result in changes to key residues in the active site of an enzyme). In spite of these complications, the notion of *structural* neutrality in a *particular* environment or genetic background remains a useful concept in the study of protein evolution.

3.2 Theory: designability

There are two, somewhat conflicting perspectives from which we can consider the relationship between genotype (sequence) and phenotype (structure). The first of these is called *designability*. We noted that the relationship mapping sequences to structures is many-to-one. What must also be observed is that the sequences are not evenly partitioned across the structures. Some structures can be generated by folding any number of a very large set of sequences; other structures are more specialized, and can only be built up from a few sequences [22]. The structures with many generative sequences are said to be more *designable* than those with fewer generative sequences. Recall that the number of protein folds observed in nature is relatively small. These folds are likely to vary amongst themselves in terms of designability; more importantly, designability is expected to vary between the observed folds and "imaginary" folds. In fact, increased designability may contribute to the dominance of the observed folds [23,24], a fact that we illustrate by example.

Let us consider a hypothetical world with two folds: one which has a useful structure, but can only be generated by a single sequence (low designability), and another which is useless, but can be generated by many sequences (high designability). If selection strongly favors utility, then clearly the first fold will propagate by virtue of its functional advantage. Designability becomes important when we introduce mutations into our model. Although the first fold has a selective advantage in its native form, it is not robust against mutations. Any change in its underlying sequence will result in a loss of its useful structural characteristics. Because the second fold is designable, it is robust against mutations, but selectively disadvantaged because it is useless. For the first fold to remain dominant, its selective advantage must be strong enough to compensate for the losses due to mutation. Now imagine a third fold, one that is both useful and designable. Selection will favor this fold like the first, because it can fill a functional role. By virtue of its designability, this fold will maintain a useful structure even while its underlying sequence accumulates mutations. All else being equal, this fold will come to dominate the population.

The example above assumed that selection acts on structure as a trait in and of itself. Hence, many mutated variants of the third fold were assumed to be selectively neutral, simply because they result in the same structure. Selection will also act on a protein's function, which may be more sensitive to specific changes in the underlying sequence. As noted above, we have employed structure as a surrogate for function, but in reality both features are important. Protein structural properties are more easily generalized than functional properties, and so the former feature tends to be more amenable to theoretical treatments.

3.3 Theory: evolvability

We can also view the genotype-phenotype (sequence-structure) relationship from the perspective of *evolvability*. Simply put, evolvability is concerned with the generation of new phenotypes from existing phenotypes—a phenomenon that is central to the evolution of species [25]. This is in stark contrast to designability, which stressed the importance of maintaining a single phenotype—avoiding change. Change, however, is at the heart of evolution. How does evolvability, and hence change, relate to the sequence-structure relationship?

Let us consider another hypothetical world for illustrative purposes. There are two dominant folds in this world, both of which result from many-to-one sequence-structure mappings. If we were to sample populations of either fold, we would find that its underlying genotypes were widely varied. We can say that these genotype sequences belong to the fold's *neutral network*—the set of all genotype sequences which produce the fold [26]. Mutating from one genotype to another within the neutral network does not change the phenotype, an example of neutral evolution. The two folds will differ in terms of their evolvabilities.

To understand what this means requires an understanding of the relationships between neutral networks. Some mutations in a given sequence result in a new sequence that remains within the neutral network of the original; other mutations result in a new sequence which belongs to a different neutral network. Mutations of the first type do not result in a change of phenotype (the fold remains constant), while mutations of the second type produce a new phenotype (the fold changes). We are interested not only in the frequency of mutations that leave a given neutral network, but also in the distribution of new networks in which they land. Do the new mutations always lead to another single network, or one of an ensemble? This information allows us to establish notions of closeness between neutral networks, and this is the essence of evolvability. A neutral network which is close to other neutral networks is evolvable—its genotypes have the potential to mutate, producing new genotypes with potentially innovative phenotypes. Conversely, an isolated neutral network is not evolvable. These ideas have been extensively tested in the context of the RNA sequence-structure relationship [27–29], in which the size of neutral networks and transition frequencies can be readily computed. The underlying theory, however, is applicable to the mapping between protein sequences and structures.

Returning to our example, assume that the neutral network of the first fold is close to those of several other folds, while the neutral network of the second fold is relatively isolated (surrounded by the neutral network of unfolded proteins, perhaps). Now assume that a change in the environment occurs causing both folds to be heavily penalized by selection. In our model, the new environment acts as an agent of selection, but does not affect the genotype-phenotype relationship (hence, the neutral networks do not change). The descendants of proteins in the second neutral network are doomed—their mutations cannot produce an innovative solution to the new environment. For the first neutral network there is hope—some of the genotypes here are likely to mutate into the neutral networks of other folds, one or more of which may fair better in the new environment.

3.4 Modeling structure and evolution

Direct observation of these theoretical forces in action is difficult due to the long timescales over which evolution operates. Ideally we would be able to model an accelerated version of this process using computers, but there are difficulties inherent to this as well. The mapping from sequence to structure (protein folding) is a spontaneous, natural process in living cells, but a major challenge in simulation. The ability to generate an accurate 3D structure of a protein by computation alone given only its amino acid sequence is the essence of the *protein folding problem* [14,30]. Despite massive amounts of work in this area, we still lack a general efficient solution.

Instead, physical models of protein evolution are usually conducted using simplified representations of proteins (such as strings of balls woven through a regular lattice) [31,32]. While these models obviously represent a gross simplification, they capture some of the physical and geometric constraints governing the sequence-structure relationship in real proteins. Results obtained in these simulations tend to be more relevant than those derived on a purely conceptual basis; the cost in terms of computational complexity is also greater. For simple models, the entire space of genotypes and phenotypes can be sampled [33], something that will likely never be possible for real proteins. Adding sophistication to these models boosts the biological relevance to their findings, but often at considerable computational costs [34–36].

Lattice models have generated a variety of interesting results, relevant to both the protein folding problem and the relationship between protein structure and evolution. The distribution of sequences in a neutral network has been explored as a function of the mutational and selective pressures on the corresponding fold [37, 38]. Similar approaches have concluded that evolution selects for sequences which can rapidly adopt their final structures [39,40]. Simulations also tend to predict small sets of dominant protein folds [24]—a result which matches our observations about the real world. The dominant simulated structures are shown to be both highly designable and thermodynamically stable, implying that a causal relationship may exist between these two quantities [18]. Whether or not these observations are generally true for real proteins is not known. The most convincing research in this area is able to pair model-based predictions with observations in a sample of real proteins. For further review of work in this area see [33].

The relationship between designability and evolvability is another area of interest currently being studied with model-based simulations [41,42]. Designable structures are advantageous because they are robust against change. Evolvable structures are advantageous because they have the ability to innovate. Do these forces oppose one another in evolution, or is there a hidden synergy between them? Lattice-based models have been used to explore the relationship between the evolution of new functions and the maintenance of stability, which is conceptually related to the designability/evolvability paradox [43,44]. While these two objectives are antagonistic under some circumstances, a period of enhanced selection for stability can promote subsequent gain of function. Other modeling approaches have also shown that evolvability is itself a selectable trait, favorable in times of rapid environmental change [45].

Clearly theories and models exploring the role of structure in protein evolution produce a wealth of fascinating ideas, many of which are supported by intuition, real world examples, and consistency with physical laws. However, observations on the role of structure in evolution which begin with real proteins in the natural world are—almost by definition—the most relevant. Our focus now turns to these observations.

4. EMPIRICAL RESULTS: SINGLE PROTEINS

4.1 Approaches

Some of the most intriguing observations about natural proteins involve the relationships between explicit physical parameters (e.g., solvent accessibility) and evolutionary rate. While it is common to think of conceptual "forces" which influence evolution, these relationships hint at true physical forces that govern the allowable changes in proteins. All of these analyses are based on real world sequence data and structures, and thus their results are highly relevant. The structure of real proteins can be determined using X-ray crystallography or NMR techniques [46,47]. These procedures are time consuming and have low throughput, but provide extremely precise (to within angstroms) 3D glimpses at the structures of real proteins. The RCSB Protein Databank, a major repository for this information, contains over 50,000 structures (as of 10/14/2008) [48]. This is a lot of data, but it pales in comparison to the millions of protein coding sequences contained in the BLAST database [49]. Therefore, another common strategy is to build models based on known structures and use these to predict physical features of the sequences whose experimental structures are not known. Both the real and inferred structural parameters can be explored at a variety of scales; from smallest to largest these include: individual residues, secondary structure motifs, protein domains, whole proteins, protein complexes, and protein networks. We reserve discussion of the last two topics for the next section.

4.2 Physical properties

One of the oldest observations linking protein structure to evolution involved the influence of solvent exposure on residue mutations. The homologous proteins hemoglobin and myoglobin, whose atomic structures had been solved by 1965, were observed to differ far more dramatically on their surfaces than in their cores [50]. This has since become a well known general feature in protein evolution. Several recent studies have reexamined the situation using large sequence and structure datasets [7,51–53]. These studies universally support the notion that buried residues in a protein's core are under tighter constraint, and therefore evolve more slowly. A protein's core—and hence, buried residues—play an important role in stabilizing its folded structure. It is therefore believed that mutations in the core may result in structure destabilization, potential misfolding, and a consequent loss of fitness. How do things change when we consider the solvent accessibility of full proteins (rather than residues in a "protein free" context)? The "functional density hypothesis," proposed in 1976, states that the selective constraint that a protein experiences should be proportional to the fraction of its residues involved with its function (e.g., catalytic activity) [54]. Although proper folding is not typically thought of as a "function" of a protein, if the protein does not fold or folds improperly, then its conventional functions will certainly be impaired. A modified "fitness density hypothesis" posits that a protein's rate of evolution should be constrained by the fraction of its residues that, if mutated, would result in a significant loss of fitness [5,11]. Buried residues would certainly seem to be among this fraction.

Intuition therefore suggests that a protein's evolutionary rate should scale with the fraction of its residues that are buried (or inversely with the fraction of solvent exposed residues). A study by Bloom and colleagues reported an opposing trend: proteins with a large fraction of buried residues seem to evolve *more* rapidly [7]. Their explanation is that proteins with large, stable cores have greater freedom to accumulate surface mutations, and that these mutations significantly elevate the overall rate at which the protein appears to be evolving. This study also considered the effect of atomic contact density on evolutionary constraint. Solvent exposure and contact density convey similar information about the three dimensional structure of a protein, and hence correlate well with one another. The Bloom et al. results regarding contact density are consistent with their solvent exposure findings: proteins with higher average contact densities appear to be evolving faster. This result is particularly interesting in light of a proposed relationship in which proteins with high contact density are also more designable [55]. Because structures with high designability have a large sequence space to explore, we might expect them to demonstrate accelerated evolution at the sequence level (as this study has found).

A subsequent study by Lin and colleagues considered both known and predicted exposure patterns in proteins with variable alignment lengths [51]. They conjecture that restricting an analysis to proteins with large alignment lengths— as was the case in the Bloom et al. study—biases results against disordered proteins, which tend to have smaller alignment lengths. Their results for proteins with smaller alignment lengths demonstrate a positive correlation between the percent of residues predicted to be solvent exposed and evolutionary rate. Results for proteins with larger alignments were consistent with the Bloom et al. findings, using either predicted or known percent exposure. The Lin et al. study makes the general observation that "proteins with a high [percentage of exposed residues] may evolve slowly or fast, whereas proteins with a low [percentage of exposed residues] almost always have a low evolutionary rate" [51]. Their conclusions stress the importance of fitness density in constraining evolutionary rate, a force which opposes designability-driven sequence divergence.

4.3 Constitutional properties

Although solvent exposure gets the most attention as a driving force in protein evolution, it is not the only physical parameter to be studied in this context. Multiple studies have considered the role of protein sequence length in evolution (which corresponds to the final size of the folded protein) [7,56]. Simple organisms have evolved with a strong evolutionary pressure for reduced genome size, which may have evolutionary implications for protein sequences [56]. A significant positive correlation between length and evolutionary rate does appear to exist, and it is much more pronounced in short proteins (less than 250 amino acids). Studying protein length also provides a good example of the complexity inherent to isolating determinants of evolution. Although length appears to correlate with evolutionary rate, it is also known to be correlated inversely with expression [56] and directly with contact density [7,56] (both of which are determinants of evolution in their own right). Carefully controlling or isolating individual factors can be a challenge, even with advanced statistical techniques.

A protein's amino acid composition should also be considered as a potential determinant of evolutionary rate. It is well known that mutations between amino acids do not occur with equal frequencies. For example, mutations that swap one hydrophobic residue for another are commonly observed, suggesting that these mutations are neutral or only slightly deleterious. In contrast, mutations between hydrophobic and hydrophilic residues are far less common, suggesting that such transitions are generally disfavored by selection. These types of observations are the basis for amino acid substitution matrices, such as PAM [57] and BLOSUM [58], which are key components of sequence alignment and other bioinformatics algorithms. Recent work has shown that the space of acceptable mutations widens with protein divergence; this result applies to both general patterns of substitution as well as the specific requirements in buried versus exposed regions [52]. An early analysis using a small set of sequences suggested that the evolutionary trajectory of a protein could be inferred based on amino acid composition alone [59]. A more recent study with a much larger dataset rejected this hypothesis, concluding that amino acid composition contributes only weakly to predictions of evolutionary rate [60]. Thus, as was the case with solvent exposure, properties at the residue level do not necessarily translate directly to whole protein behavior. As far as evolution is concerned, proteins appear to be more than just a sum of their parts.

The next level up in protein organization involves secondary structures motifs—small structural elements that show up repeatedly within many different protein folds. Well-known examples include helices, strands, loops, and turns. Work in yeast has shown that the secondary structure composition of a protein does not appear to influence its evolutionary rate [7]. However, in a study of mammalian proteins, residues in helices and strands were shown to evolve more slowly than those in the less ordered loops and turns [53]. This last result highlights another apparent influence on evolution: molecular disorder. Disordered regions of proteins are generally known to evolve more rapidly than their ordered counterparts [61]. Our discussion to this point has assumed that a useful structure and a stable fold are synonymous—this is not necessarily the case. In fact, many proteins perform functional roles in the cell despite the fact that they, either in whole or in part, fail to achieve a fixed three-dimensional fold [62]. The fact that these proteins also appear to experience relaxed selection raises interesting questions about their evolutionary potential.

4.4 Protein domains

Protein domains, the next level up in the hierarchy of protein constitution, are important enough to warrant a separate discussion. Domains can be defined based on function, structure, or sequence characterization; in many cases the different approaches are compatible. We naturally adopt a structure-based definition: a protein domain is a spatially distinct structure (or structural component) that could conceivably fold and function in isolation [63]. Some proteins consist of a single domain, while others are composed of multiple domains each folded separately from a subsection of the underlying amino acid chain. To this point, our notion of the genotype to phenotype relationship has been protein sequence → protein structure. Given the discrete spatial nature of domains, protein *sub*sequence → domain would be an equally valid definition. In fact, the notion of a protein fold (as in, "this fold is highly designable") translates naturally to the protein domain concept.

To our knowledge, "domain constitution" of a protein has not been considered as a determinant of evolutionary rate. This results from the fact that domains are large, discrete units of proteins, unlike fine scale properties like buried residues or secondary structure elements. Instead, domains are typically considered as evolutionary targets in their own right [64]. Domains share a natural history that is similar in many ways to the phylogenies describing evolution at the level of whole organisms [63]. Many modern domains are thought to have evolved and radiated from lineages of ancestral domains, which were in turn derived from primordial protein folds. Domains without a shared evolutionary history may have also acquired similar structures due to convergent evolution [65]. Note that this is highly compatible with notions from designability theory. Classifying domains based on these principles is the primary mission of databases such as Pfam [17], CATH [66], and SCOP [15]. Some relationships between domains can be inferred at the sequence level, but owing to the many-to-one mapping of sequences to structures, structure comparison methods are often critical for describing connections between domains from distantly diverged proteins.

The discrete nature of domains has played an important role in protein evolution. After the evolution of a handful of primordial domains, many new functions could be efficiently evolved through their combination and permutation [67]. This process is facilitated by genomic evolution, in which pieces of genetic material (such as those encoding amino acid subsequences responsible for protein domains) are readily duplicated, fused, shuffled, and deleted [68]. In cases of domain duplication, while the original template continues to fill its role in the cell, the duplicate has the freedom to explore sequence and structure space, possibly acquiring new functions in the process [64]. The distribution of domains and proteins produced in this process follows power law behavior [69,70], which is emerging as a common trait among large scale biological systems. While interesting, genomic perspectives on domain evolution take us too far afield from our structural focus.

4.5 Function

As mentioned in the introduction, it has been suggested that a protein's functional classification is generally not a good predictor of its evolutionary rate [6]. However, some basic functional attributes of a protein certainly have important structural and evolutionary implications. For example, Kimura and Ohta demonstrated as far back as 1973 that residues involved with binding the heme group in α and β globin (the protein constituents of the hemoglobin molecule) evolve at one tenth the rate of the background structure [71]. Residues like these contribute to a protein's functional density and hence to the revised fitness density as well. Unlike structural properties that are common to many (or all) proteins, structure-function properties tend to be highly specialized, and are better reviewed on a case-by-case basis. A great deal of literature linking specific structure-function relationships to evolution is available for the interested reader.

5. EMPIRICAL RESULTS: HIGHER ORDER PROPERTIES

5.1 Interfaces

We begin our discussion of higher order structural properties with a final single protein property: interfaces. Although interfaces are properties of the unique protein structure to which they belong, they form a variety of interesting larger structures—each with evolutionary significance—when we consider them together. Interfaces are intimately linked with the notions of solvent accessibility and burial discussed previously, and several studies have investigated both simultaneously. We consider this to be a preferred approach, as interfacial residues and surface area (and their evolutionary contributions) will be wrongly counted as exposed residues and surface area when proteins are considered independently.

An early study of the cytochrome c protein structure revealed that some portions of the surface seemed to be experiencing unusually high functional constraint [72]. These surface residues were determined to be sites of interaction with other proteins (interfaces). Subsequent studies have generally supported the notion that interfacial surfaces are more conserved than the remainder of the protein's solvent-exposed surface, and slightly less conserved than the protein's core [73]. Substitutions that *do* occur in the interface are heavily skewed toward more conservative changes [53], as defined by the Grantham classification scheme [74]. Exploiting the difference in evolutionary rate between interfacial and non-interfacial sections of a protein's surface has been proposed as a means by which to identify interfaces in newly characterized proteins; this has proven to be difficult in practice [75].

The notion that evolutionary rate of an "average interface" is intermediate to those of buried and solvent-exposed portions of a protein seems very intuitive. Interfaces will likely spend at part of their lives in a buried state (when interacting) and another part in a solvent-exposed state (when not interacting). One might therefore expect the rate of evolution at an interface to scale inversely with the proportion of time that it is active; indeed, this is precisely what has been found [76].

Evolutionary rate among residues belonging to transient interfaces is significantly higher than for those found in constitutive (permanent) interfaces; rates for both are intermediate to those of buried and solvent-exposed residues. Decreased evolutionary rate at constitutive interaction sites may also reflect specific structural constraints imposed by the protein's interaction partner [76]. This represents a case of *coevolution* between protein structures, an instance of a higher order structure-evolution relationship.

5.2 Protein–protein interaction networks

Studying the topological structure (not to be confused with molecular structure) of protein–protein interaction networks is a hot topic in systems biology research. In such a network, proteins are represented as vertices, and interactions between protein pairs are represented as edges. For our purposes, interactions can be thought of as direct physical connections between the involved proteins (such as those mediated by interfaces); other common notions of protein–protein interactions exist that are equally important, but they lack structural significance. Interactions in these networks tend to follow a power law distribution, such that a small number of proteins have a very high degree (many interactions) and a large number have a very low degree (few interactions) [77]. Proteins with many interaction partners (hubs) tend to be essential and evolve slowly; whether or not there is a functional dependence between the number of a protein's interaction partners and its selective constraint has been a topic of contention [78]. Integrating network topology with expression data has also shown that hub proteins can be divided into two classes based on the timing of their interactions:

(1) *party hubs*, which interact with several partners simultaneously, and
(2) *date hubs*, which interact in a "one-partner-at-a-time" fashion [79].

The importance of integrating structural information into biological networks has been recognized [80], but relatively few studies have actually taken this leap. One such study related the number and extent of interfacial surfaces on a protein to its behavior in a network [81]. This approach allowed hubs to be partitioned into multi-interface and singlish-interface classes, which act as structural analogs of the temporal party and date hub classifications, respectively (note: *singlish* implies 1 or 2 interfaces). While singlish-interface hubs can evolve a new interaction through duplication and divergence of a partner, new interactions in multi-interface hubs necessitate the creation of a new binding interface. Finally, the study concludes that the extent of a protein's surface area involved in interactions is a better predictor of evolutionary rate than its number of interaction partners, in agreement with previous proposals [82]. Other efforts have directly employed structural information in network construction. We mentioned previously that it is difficult to predict interfacial components of a protein's surface based on conservation alone. Another structural approach to interaction prediction involves the consideration of protein domains [83]. Recall that domains are large subsections of proteins, typically with well conserved, discrete structures. Imagine that Domain A in Protein 1 and

Domain *B* in Protein 2 are found to physically interact. Protein 3, having uncharacterized interaction potential, is found to contain Domain *B*, either by structure or sequence comparison. It is reasonable to hypothesize that an interaction between Protein 1 and Protein 3 may occur in the cell. Networks based on domain interactions have been created following this logic with great success [84,85]. These networks further highlight the importance of structural modularity in the evolution of single proteins and protein networks. Returning to the example, while it is *possible* for Proteins 1 and 3 to interact on the basis of conserved domain relationships, this interaction is not a given. Conserved domain pairs that violate this assumption are common, and typically only differ by a few surface mutations; these subtle changes are enough to dramatically decrease the stability of the interaction [86].

Disordered (unfolded) regions of a protein are known to perform important biological functions, in spite of relaxed constraint on their three dimensional structures. It has been shown that hub proteins, which are believed to be constrained by coevolution with their interaction partners, are also more likely to feature intrinsic disorder [87]. This provides another example of a pair of deterministic forces in evolution with a paradoxical relationship. A recent work by Kim et al. addresses this issue by considering the precise physical context of disordered regions that occur in interacting proteins [88].

5.3 Protein complexes

Proteins seldom perform their functions in isolation. Either for the purpose of building multi-component architectural structures or streamlining functions, proteins are often grouped in space as complexes. This higher level structure is metaphorically similar to the way in which domains are grouped to build more sophisticated individual proteins. Note however that the combination of discrete proteins into complexes is a purely physical process, whereas domains are linked both physically and genetically by the underlying protein sequence. This first definition of a protein complex is generally given to groups of proteins which all interact constitutively. Complexes in this sense are analogs of the party hubs in expression-based networks and multi-interface hubs in structure-based networks. Evolutionary insights about these hubs are equally applicable to complexes, and vice versa.

One of the interesting connections between evolution and structure in complexes relates to the *balance hypothesis*, as described by Papp et al. [89]. This hypothesis states that changes which affect the proportions of complex-forming proteins in a cell will be deleterious, and hence purged by selection. A reduction in availability (either whole or partial) of a complex component limits the number of complete complexes that the cell can build, which may have obvious fitness consequences. Perhaps less intuitive is the fact that an over-available component may also represent a fitness loss, either by disrupting the kinetics of proper structure assembly, or by carrying out some "unsupervised" activity in its lone state. Thus, evolution acts to maintain fixed stoichiometry among the components of important complexes. It is possible, however, for the genomic segment encoding the entire complex to be duplicated, as in this case the stoichiometry among complex components is maintained.

6. SUMMATION

The role that structure plays in protein evolution is evident on many scales. Single residues feel differences in selective constraint according to the extent of their solvent exposure. Whole proteins have a freedom to diverge that varies with the degree of disorder in their native structures. Cassettes of independent structures evolve together in order to maintain strict interaction proportions. These are examples of observations that have been made by considering real protein structures. Above this level there exists an armamentarium of theory and models to describe the structurally significant trends or events in protein evolution that we have not yet been able to observe directly. Much has been learned, and much remains to be discovered. Several ideas will motivate future work at the interface of structural and evolutionary biology.

(i) **More atomic level structure data is needed.** While few would scoff at the set of structures available in current databases, this represents only a minute fraction of the proteins present in nature today. We have a notion that a small set of structures is likely to be dominant among the protein universe. This notion lends itself well to summary and classification, but not to exhaustive description of the protein universe. As we saw in the case of domain interactions, the difference of a few amino acids in otherwise identical folds can be enough to significantly differentiate their behaviors. Advances in structure determination methods will be useful for increasing the pool of solved lone and complexed protein structures, which can then be pipelined into theoretical and empirical studies.

(ii) **New ways of considering structure must be developed.** Protein structures mediate biological function, and their evolution is shaped by that relationship. Approaches that integrate structural information into biological analysis, particularly analysis at the systems scale, will produce a more complete picture of the mechanisms that drive living organisms. Also implicit in this idea is a need for new methods to describe structures. Designability, evolvability, and fitness density are significant quantitative structural measures that influence a protein's evolution. How to precisely define and determine quantities such as these in real proteins remains an open question.

(iii) **Theoretical and empirical results must keep pace with one another.** We have a wealth of theoretical ideas concerning structure-evolution relationships. While many of these are very intuitive and have a "sense" of relevance, true relevance must be earned through explorations in real world systems. Advances in modeling of the sequence-structure relationship—e.g., progress made in the protein folding problem—will facilitate more realistic *in silico* models of protein evolution. Better integration of available data and directed laboratory evolution experiments will also aid in this goal. From the opposite perspective, theoretical treatments of structural (and other) determinants of protein evolution must be advanced to better handle the wealth of data available to them. Thus far, the inherent noise and vast interconnections among these biological variables have proven to be worthy adversaries to state-of-the-art analysis methods.

Current work in protein design and directed evolution promises to produce exciting new discoveries at the interface of structure and evolution. These approaches seek to simulate the evolutionary processes we have discussed here in a laboratory context, providing researchers with a realistically short evolutionary timescale (the advantage of theoretical work) and the relevance of working with real proteins (the advantage of empirical observation). For further review of these approaches, see [90,91].

With continued progress toward these research objectives, we expect that our knowledge of biology and evolution will continue to strengthen in the light of molecular structure.

ACKNOWLEDGMENTS

Y.X. is supported by a Research Starter Grant in Informatics from the PhRMA Foundation. E.F. is supported by an IGERT Fellowship through NSF grant DGE-0654108 awarded to the BU Bioinformatics Program.

REFERENCES

1. Dobzhansky, T. Biology, molecular and organismic. Am. Zool. 1964, 4, 443–52.
2. McInerney, J.O. The causes of protein evolutionary rate variation. Trends Ecol. Evol. 2006, 21(5), 230–2.
3. Kellis, M., Birren, B.W., Lander, E.S. Proof and evolutionary analysis of ancient genome duplication in the yeast Saccharomyces cerevisiae. Nature 2004, 428(6983), 617–24.
4. Hurst, L.D. The Ka/Ks ratio: Diagnosing the form of sequence evolution. Trends Genet. 2002, 18(9), 486.
5. Pal, C., Papp, B., Lercher, M.J. An integrated view of protein evolution. Nat. Rev. Genet. 2006, 7(5), 337–48.
6. Rocha, E.P. The quest for the universals of protein evolution. Trends Genet. 2006, 22(8), 412–6.
7. Bloom, J.D., et al. Structural determinants of the rate of protein evolution in yeast. Mol. Biol. Evol. 2006, 23(9), 1751–61.
8. Plotkin, J.B., Fraser, H.B. Assessing the determinants of evolutionary rates in the presence of noise. Mol. Biol. Evol. 2007, 24(5), 1113–21.
9. Rocha, E.P., Danchin, A. An analysis of determinants of amino acids substitution rates in bacterial proteins. Mol. Biol. Evol. 2004, 21(1), 108–16.
10. Akashi, H. Translational selection and yeast proteome evolution. Genetics 2003, 164(4), 1291–303.
11. Drummond, D.A., et al. Why highly expressed proteins evolve slowly. Proc. Natl. Acad. Sci. USA 2005, 102(40), 14338–43.
12. Goldberg, A.L. Protein degradation and protection against misfolded or damaged proteins. Nature 2003, 426(6968), 895–9.
13. Crick, F. Central dogma of molecular biology. Nature 1970, 227(5258), 561–3.
14. Dill, K.A., et al. The protein folding problem: When will it be solved?. Curr. Opin. Struct. Biol. 2007, 17(3), 342–6.
15. Hubbard, T.J., et al. SCOP: A structural classification of proteins database. Nucleic Acids Res. 1997, 25(1), 236–9.
16. Chothia, C. Proteins. One thousand families for the molecular biologist. Nature 1992, 357(6379), 543–4.

17. Bateman, A., et al. The Pfam protein families database. Nucleic Acids Res. 2002, 30(1), 276–80.
18. Helling, R., et al. The designability of protein structures. J. Mol. Graph Model 2001, 19(1), 157–67.
19. Kimura, M. The Neutral Theory of Molecular Evolution Cambridge: Cambridge Univ. Press; 1983.
20. Wagner, A. Robustness, evolvability, and neutrality. FEBS Lett. 2005, 579(8), 1772–8.
21. Sharp, P.M., et al. DNA sequence evolution: The sounds of silence. Philos. Trans. R. Soc. Lond. B Biol. Sci. 1995, 349(1329), 241–7.
22. Wong, P., Frishman, D. Fold designability, distribution, and disease. PLoS Comput. Biol. 2006, 2(5), e40.
23. Li, H., Tang, C., Wingreen, N.S. Are protein folds atypical?. Proc. Natl. Acad. Sci. USA 1998, 95(9), 4987–90.
24. Li, H., et al. Emergence of preferred structures in a simple model of protein folding. Science 1996, 273(5275), 666–9.
25. Kirschner, M., Gerhart, J. Evolvability. Proc. Natl. Acad. Sci. USA 1998, 95(15), 8420–7.
26. Huynen, M.A., Stadler, P.F., Fontana, W. Smoothness within ruggedness: The role of neutrality in adaptation. Proc. Natl. Acad. Sci. USA 1996, 93(1), 397–401.
27. Stadler, B.M., et al. The topology of the possible: Formal spaces underlying patterns of evolutionary change. J. Theor. Biol. 2001, 213(2), 241–74.
28. Fontana, W., Schuster, P. Shaping space: The possible and the attainable in RNA genotype-phenotype mapping. J. Theor. Biol. 1998, 194(4), 491–515.
29. Schuster, P., et al. From sequences to shapes and back: A case study in RNA secondary structures. Proc. Biol. Sci. 1994, 255(1344), 279–84.
30. Itzhaki, L., Wolynes, P. The quest to understand protein folding. Curr. Opin. Struct. Biol. 2008, 18(1), 1–3.
31. Dill, K.A., et al. Principles of protein folding—A perspective from simple exact models. Protein Sci. 1995, 4(4), 561–602.
32. Shakhnovich, E.I. Theoretical studies of protein-folding thermodynamics and kinetics. Curr. Opin. Struct. Biol. 1997, 7(1), 29–40.
33. Xia, Y., Levitt, M. Simulating protein evolution in sequence and structure space. Curr. Opin. Struct. Biol. 2004, 14(2), 202–7.
34. Hinds, D.A., Levitt, M. From structure to sequence and back again. J. Mol. Biol. 1996, 258(1), 201–9.
35. Park, B.H., Levitt, M. The complexity and accuracy of discrete state models of protein structure. J. Mol. Biol. 1995, 249(2), 493–507.
36. Mirny, L., Shakhnovich, E. Protein folding theory: From lattice to all-atom models. Ann. Rev. Biophys. Biomol. Struct. 2001, 30, 361–96.
37. Xia, Y., Levitt, M. Roles of mutation and recombination in the evolution of protein thermodynamics. Proc. Natl. Acad. Sci. USA 2002, 99(16), 10382–7.
38. Xia, Y., Levitt, M. Funnel-like organization in sequence space determines the distributions of protein stability and folding rate preferred by evolution. Proteins 2004, 55(1), 107–14.
39. Mirny, L.A., Abkevich, V.I., Shakhnovich, E.I. How evolution makes proteins fold quickly. Proc. Natl. Acad. Sci. USA 1998, 95(9), 4976–81.
40. Gutin, A.M., Abkevich, V.I., Shakhnovich, E.I. Evolution-like selection of fast-folding model proteins. Proc. Natl. Acad. Sci. USA 1995, 92(5), 1282–6.
41. Wagner, A. Robustness and evolvability: A paradox resolved. Proc. Biol. Sci. 2008, 275(1630), 91–100.
42. Lenski, R.E., Barrick, J.E., Ofria, C. Balancing robustness and evolvability. PLoS Biol. 2006, 4(12), e428.
43. Bloom, J.D., et al. Protein stability promotes evolvability. Proc. Natl. Acad. Sci. USA 2006, 103(15), 5869–74.
44. Bloom, J.D., et al. Stability and the evolvability of function in a model protein. Biophys. J. 2004, 86(5), 2758–64.
45. Earl, D.J., Deem, M.W. Evolvability is a selectable trait. Proc. Natl. Acad. Sci. USA 2004, 101(32), 11531–6.
46. Drenth, J. Principles of X-Ray Crystallography New York: Springer; 1999.
47. Clore, G.M., Gronenborn, A.M. Determining the structures of large proteins and protein complexes by NMR. Trends Biotechnol. 1998, 16(1), 22–34.

48. Bernstein, F.C., et al. The Protein Data Bank: A computer-based archival file for macromolecular structures. J. Mol. Biol. 1977, 112(3), 535–42.
49. Altschul, S.F., et al. Gapped BLAST and PSI-BLAST: A new generation of protein database search programs. Nucleic Acids Res. 1997, 25(17), 3389–402.
50. Perutz, M.F., Kendrew, J.C., Watson, H.C. Structure and function of haemoglobin II. Some relations between polypeptide chain configuration and amino acid sequence. J. Mol. Biol. 1965, 13, 669–78.
51. Lin, Y.S., et al. Proportion of solvent-exposed amino acids in a protein and rate of protein evolution. Mol. Biol. Evol. 2007, 24(4), 1005–11.
52. Sasidharan, R., Chothia, C. The selection of acceptable protein mutations. Proc. Natl. Acad. Sci. USA 2007, 104(24), 10080–5.
53. Choi, S.S., Vallender, E.J., Lahn, B.T. Systematically assessing the influence of 3-dimensional structural context on the molecular evolution of mammalian proteomes. Mol. Biol. Evol. 2006, 23(11), 2131–3.
54. Zuckerkandl, E. Evolutionary processes and evolutionary noise at the molecular level. I. Functional density in proteins. J. Mol. Evol. 1976, 7(3), 167–83.
55. England, J.L., Shakhnovich, E.I. Structural determinant of protein designability. Phys. Rev. Lett. 2003, 90(21), 218101.
56. Warringer, J., Blomberg, A. Evolutionary constraints on yeast protein size. BMC Evol. Biol. 2006, 6, 61.
57. Dayhoff, M., Schwartz, R.M., Orcutt, B. Atlas of Protein Sequence and Structure Silver Spring: National Biomedical Research Foundation; 1978.
58. Henikoff, S., Henikoff, J.G. Amino acid substitution matrices from protein blocks. Proc. Natl. Acad. Sci. USA 1992, 89(22), 10915–9.
59. Graur, D. Amino acid composition and the evolutionary rates of protein-coding genes. J. Mol. Evol. 1985, 22(1), 53–62.
60. Tourasse, N.J., Li, W.H. Selective constraints, amino acid composition, and the rate of protein evolution. Mol. Biol. Evol. 2000, 17(4), 656–64.
61. Brown, C.J., et al. Evolutionary rate heterogeneity in proteins with long disordered regions. J. Mol. Evol. 2002, 55(1), 104–10.
62. Wright, P.E., Dyson, H.J. Intrinsically unstructured proteins: Re-assessing the protein structure-function paradigm. J. Mol. Biol. 1999, 293(2), 321–31.
63. Ponting, C.P., Russell, R.R. The natural history of protein domains. Ann. Rev. Biophys. Biomol. Struct. 2002, 31, 45–71.
64. Orengo, C.A., Thornton, J.M. Protein families and their evolution—A structural perspective. Ann. Rev. Biochem. 2005, 74, 867–900.
65. Lupas, A.N., Ponting, C.P., Russell, R.B. On the evolution of protein folds: Are similar motifs in different protein folds the result of convergence, insertion, or relics of an ancient peptide world? J. Struct. Biol. 2001, 134(2–3), 191–203.
66. Orengo, C.A., et al. CATH—A hierarchic classification of protein domain structures. Structure 1997, 5(8), 1093–108.
67. Chothia, C., et al. Evolution of the protein repertoire. Science 2003, 300(5626), 1701–3.
68. Bjorklund, A.K., et al. Domain rearrangements in protein evolution. J. Mol. Biol. 2005, 353(4), 911–23.
69. Zhang, C., DeLisi, C. Estimating the number of protein folds. J. Mol. Biol. 1998, 284(5), 1301–5.
70. Dokholyan, N.V., Shakhnovich, B., Shakhnovich, E.I. Expanding protein universe and its origin from the biological Big Bang. Proc. Natl. Acad. Sci. USA 2002, 99(22), 14132–6.
71. Kimura, M., Ota, T. Mutation and evolution at the molecular level. Genetics 1973, 73(Suppl 73), 19–35.
72. Dickerson, R.E. The structures of cytochrome c and the rates of molecular evolution. J. Mol. Evol. 1971, 1(1), 26–45.
73. Valdar, W.S., Thornton, J.M. Protein–protein interfaces: Analysis of amino acid conservation in homodimers. Proteins 2001, 42(1), 108–24.
74. Grantham, R. Amino acid difference formula to help explain protein evolution. Science 1974, 185(4154), 862–4.
75. Caffrey, D.R., et al. Are protein–protein interfaces more conserved in sequence than the rest of the protein surface? Protein Sci. 2004, 13(1), 190–202.

76. Mintseris, J., Weng, Z. Structure, function, and evolution of transient and obligate protein–protein interactions. Proc. Natl. Acad. Sci. USA 2005, 102(31), 10930–5.
77. Albert, R. Scale-free networks in cell biology. J. Cell. Sci. 2005, 118(Pt 21), 4947–57.
78. Jordan, I.K., Wolf, Y.I., Koonin, E.V. No simple dependence between protein evolution rate and the number of protein–protein interactions: Only the most prolific interactors tend to evolve slowly. BMC Evol. Biol. 2003, 3, 1.
79. Han, J.D., et al. Evidence for dynamically organized modularity in the yeast protein–protein interaction network. Nature 2004, 430(6995), 88–93.
80. Aloy, P., Russell, R.B. Structural systems biology: Modelling protein interactions. Nat. Rev. Mol. Cell. Biol. 2006, 7(3), 188–97.
81. Kim, P.M., et al. Relating three-dimensional structures to protein networks provides evolutionary insights. Science 2006, 314(5807), 1938–41.
82. Fraser, H.B., et al. Evolutionary rate in the protein interaction network. Science 2002, 296(5568), 750–2.
83. Kiel, C., Beltrao, P., Serrano, L. Analyzing protein interaction networks using structural information. Ann. Rev. Biochem. 2008, 0, 0.
84. Deng, M., et al. Inferring domain–domain interactions from protein–protein interactions. Genome Res. 2002, 12(10), 1540–8.
85. Schlicker, A., et al. Functional evaluation of domain–domain interactions and human protein interaction networks. Bioinformatics 2007, 23(7), 859–65.
86. Kiel, C., Serrano, L. Prediction of Ras-effector interactions using position energy matrices. Bioinformatics 2007, 23(17), 2226–30.
87. Haynes, C., et al. Intrinsic disorder is a common feature of hub proteins from four eukaryotic interactomes. PLoS Comput. Biol. 2006, 2(8), e100.
88. Kim, P.M., et al. The role of disorder in interaction networks: A structural analysis. Mol. Syst. Biol. 2008, 4, 179.
89. Papp, B., Pal, C., Hurst, L.D. Dosage sensitivity and the evolution of gene families in yeast. Nature 2003, 424(6945), 194–7.
90. Pokala, N., Handel, T.M. Review: protein design—Where we were, where we are, where we're going. J. Struct. Biol. 2001, 134(2–3), 269–81.
91. Farinas, E.T., Bulter, T., Arnold, F.H. Directed enzyme evolution. Curr. Opin. Biotechnol. 2001, 12(6), 545–51.

CHAPTER 2

Predicting Selectivity and Druggability in Drug Discovery

Alan C. Cheng*

Contents

1. INTRODUCTION

Druggability and selectivity analysis are increasingly performed in early drug discovery for both target assessment and setting lead optimization strategies. This is necessitated by the high failure rates in the drug discovery process—greater than 60% in early drug discovery screening and lead optimization stages alone [1]. In target assessment, ideas are rated on target-validation, assay feasibility, druggability, and selectivity as it relates to toxicity and side-effect potential [2,3]. In lead optimization, selectivity analysis can suggest both possible selectivity issues, as well as regions of the binding site that allow the drug discovery team to overcome these issues. Druggability analysis can be useful in suggesting additional "hot spots" for increasing potency of lead compounds. This review covers computational approaches for assessing and predicting selectivity and druggability, and those interested in computational aspects of target validation may want to read a recent review by Loging et al. [4].

* Amgen Inc., One Kendall Square, Bldg 1000, Cambridge, MA 02139, USA. E-mail: alan.cheng@amgen.com

Annual Reports in Computational Chemistry, Vol. 4
ISSN 1574-1400, DOI: 10.1016/S1574-1400(08)00002-9

2. SELECTIVITY

Selectivity analysis has the goal of identifying potential secondary pharmacology and suggesting assays for following up predicted selectivity issues, as well as identifying strategies for improving selectivity of small molecule leads. Profiling for and optimizing against secondary pharmacology is important to the discovery of compounds with decreased side effects, more desirable therapeutic profiles, and greater therapeutic differentiation. For example, successful kinase inhibitors such as imantinib (Gleevec) and sutanimib (Sutent) have a distinct selectivity profile that confers efficacy and safety [5]. In terms of computational approaches, analyses of increasing sophistication can be performed depending on how much information is available—this includes protein sequence, protein structure, and ligand information. The wide availability of these types of information for protein kinases has allowed for a significant body of selectivity analysis work, which is covered in reviews such as [6–9]. Readers interested in off-target cytochrome P450 inhibition, transporter-mediated efflux, and ADMET-related prediction for small molecules may want to consider recent reviews in the Annual Reports in Computational Chemistry [10–12].

2.1 Ligand analysis

If ligands are known for the biological target, cheminformatics approaches are useful in identifying potential selectivity issues, especially when they are used in combination with aggregate compound databases that are annotated with biological activity. Such databases include historical databases maintained in-house at biopharmaceutical companies, as well as Jubilant, GVK, MDDR, WOMBAT, and StARLITe databases [13,14]. The idea of comparing targets by looking at the small molecules that modulate them has been termed SARAH, for structure-activity relationship homology [15]. In the original SARAH idea, experimentally measured affinities for a diverse set of compounds represent an "affinity fingerprint" for a target, and similar pharmacological profiles would indicate target homology in SAR space [15]. From a drug discovery perspective, this approach is meaningful since it identifies similarity based on inhibitor or antagonist/agonist profiles, and an example where this experimental approach is applied to cysteine protease inhibitors is described in [16].

Small molecule screening data accumulated in compound databases can also be used in identifying target homology in SAR space. One approach is to identify analogs of the known active compounds using similarity searches based on 2D chemical fingerprints [17], and then look at the biological activities of the identified compounds. Conceptually, the confidence in the predicted selectivity issue increases as a pair of biological targets share larger numbers of chemotypes. This approach has traditionally been qualitative and subjective, and recently several groups have sought to make the approach more systematic and rigorous [18]. One way, termed the similarity ensemble approach (SEA) [19], takes the summed similarity score over all pairs of ligands that two biological targets share and compares it to the distribution of scores from random sets of compounds, thus allowing calculation of a statistical confidence value that is similar to the E-value used in scoring BLAST [20] sequence searches. Another approach is to generate a similarity score between two sets of ligands, but use Bayesian models to weight compound substructures that contribute more to activity [21]. A pure machine learning approach can also be used, and involves training activity models and then using the models to predict off-target activities. Nidhi et al. used a naïve Bayesian classifier to train models for 964 biological target activities and found 77% prediction accuracy when predicting activities for a separate data set [22].

These chemo-centric similarity approaches can help in identifying a selectivity panel if there is sufficient *a priori* data, and can also be used after a high-throughput screen (HTS) is complete and more ligands are known. The database SARAH approach has been applied in varying degrees of sophistication to nuclear hormone receptors [23], kinases [8], and enzymes in general [24].

2.2 Sequence analysis

While ligand information is not always available, protein sequence information is almost always available. Starting with the protein sequence, related proteins, or homologs, can be found through sequence similarity searches such as BLAST [20], where the typical search is a protein BLAST against human sequences in the non-redundant sequence database [25]. Rat and mouse sequences may also be of interest depending on the disease model that will be used. Once protein sequences are identified, multiple sequence alignment of significant BLAST hits can be performed using programs such as Clustal [26].

A variety of methods are available to cluster sequences and identify similarities starting from the multiple sequence alignment, with the most straightforward of these being pair-wise measurement of sequence identity or sequence similarity. Sequence identity is the percent of the residue positions that match, while sequence similarity involves a substitution matrix where amino acid residue similarity is taken into account. A more sophisticated approach to identify significantly related proteins is to infer a phylogenetic tree based on the multiple sequence alignment. Closely related proteins in a phylogenetic tree are, in general, likely to be selectivity issues. Similarity and phylogenetic tree calculations can be performed using software tools such as PFAAT [27], Jalview [28], and Mega [29], which are listed in Table 2.1.

When information on the protein domain of interest is available, the sequence analysis can be focused on the domain sequences. An even more detailed investigation of residues around the binding site can me made if there is information about the desired drug interaction site from experimental data such as mutagenesis results or co-crystal structure information. For instance, in analyzing kinase selectivity issues, workers often focus on residues lining the ATP binding site [30,31]. In these binding site analyses, it is important to note that all residues do not contribute equally to binding, and close inspection of the actual interactions based on a crystal structure would be prudent [32,33]. For instance, a protein backbone interaction to the ligand does not depend strongly on amino acid type, and prolines can change or rigidify the main chain conformation. An analysis of ki-

Table 2.1 Some popular tools for performing selectivity analysis using protein sequence

Task	Tool/resource	Web link
Search for related protein sequences	Blast	http://www.ncbi.nlm.nih.gov/blast
Multiple sequence alignment	Clustal	http://bips.u-strasbg.fr/en/Documentation/ClustalX/
Analyze a multiple sequence alignment	PFAAT Jalview Mega	http://pfaat.sourceforge.net http://www.jalview.org http://www.megasoftware.net
Identifying domains	NCBI conserved domain database	http://www.ncbi.nlm.nih.gov/Structure/cdd/

nase inhibitors found that two non-conservative, energetically-important, residue substitutions in the binding site are sufficient for gaining selectivity for a compound [34].

When co-crystal structures are not available, Ortiz et al. have suggested using functional residue prediction methods to identify selectivity residues [35]. The most popular of these methods are Evolutionary Trace [36] and ConSurf [37], which use phylogenetic trees to predict biologically-relevant residues that are then mapped onto a representative crystal structure.

2.3 Structure-based analysis

One significant limitation of sequence-based approaches is the inability to assess selectivity issues between targets lacking sequence homology. For instance, protein kinase sequences cannot be aligned to phosphodiesterase sequences even though selectivity issues have been observed between the two target classes. Structure-based approaches can help to identify non-homologous selectivity concerns when co-crystal structures are available. Such approaches can be classified by whether they are receptor-focused or ligand-focused, as described below.

Receptor-focused approaches involve comparison of the physiochemical properties of residues that line the binding pocket. CavBase [38], SURFACE [39], and SitesBase [40] are three examples. CavBase converts the portions of binding site residues exposed to solvent into sets of points defined by one of five pseudocenter types (aliphatic, donor, acceptor, donor/acceptor, and aromatic). For instance, a tyrosine is represented by an "aromatic" pseudocenter placed in the middle of the phenyl ring and a "donor/acceptor" pseudocenter placed at the oxygen of the hydroxyl. The constellation of pseudocenters representing a binding site is then compared to those of other binding sites using clique-detection algorithms that identify matching portions of the constellations. This approach has been applied to classification of the enzyme binding pocket in protein kinases [41]. SURFACE represents each residue using just two pseudocenters, which represents the backbone $C\alpha$ atom and the side-chain center of mass [39]. Instead of just scoring for matches, an evolutionary amino acid substitution matrix is used. SitesBase compares binding sites based on actual atoms and atom types (carbon, nitrogen, oxygen, sulfur) instead of pseudocenters [40], and the approach was applied to proteases [42].

Instead of basing comparisons off properties of residues that flank the binding site, ligand-focused approaches attempt to compare the actual small molecule binding space. GRID/PCA [43,44], for instance, uses GRID to systematically sample the binding site with a set of chemical probes, and uses an energy function to generate molecular interaction fields that represent areas of favorable affinity for each of the probes. Applying principle component analysis (PCA) to the GRID values then identifies consistency as well as differences in the interaction fields. The method has been used to study a set of 13 ephrin receptor tyrosine kinases [45] as well as a set of ten structures of CDK2 and GSK-3b [46]. Reported applications of GRID/PCA have generally focused on selectivity issues among proteins with sequence homology, in part due to the necessity of receptor structure superposition. The more recent GRIND/PCA method [43] does allow for comparisons independent of structural alignment, and has been used to compare a homology model of adenosine receptor A1 to four ribose-binding proteins [38]. The small number of comparisons is likely due to the compute-intensive nature of GRID calculations. Another approach [47] uses docking to identify a set of predicted active compounds for the protein of interest, and this set is then docked to possible selectivity targets. Targets with the most similar binding sites were shown to have the highest docking scores.

For lead optimization, Sheinerman et al. [34] showed in the context of protein kinases that energetically important residues could be identified using a systematic analysis of small-molecule structure-activity relationships in the context of a protein family sequence alignment and available structures for compound binding modes. A quantitative method for optimization of electrostatic interactions—including accounting for desolvation effects—was demonstrated for HIV protease inhibitor design recently [48]. The approach uses mathematical optimization techniques to define a ligand with maximal potency against a desirable set of targets (a set of escape mutants for HIV protease) and minimal potency for an undesirable set of human aspartyl protease "decoys" [48]. A rigorous but decidedly theoretical biophysical inquiry into the physical basis of selectivity found that polar and charged groups increase specificity of ligand interactions due to their greater sensitivity to shape complementarity as compared to hydrophobic interactions, and, in addition, conformational flexibility can increase the specificity of polar and charged interactions [49].

Molecular analyses are imperfect in that serendipitous binding modes are always possible. For instance, crystal structures of PXR show that ligands have multiple binding modes [50], and protein kinases can adopt multiple inactive conformations that are druggable [51]. A dramatic example is shown in Figure 2.1, where a small modification to a non-selective kinase inhibitor yielded $1400\times$ se-

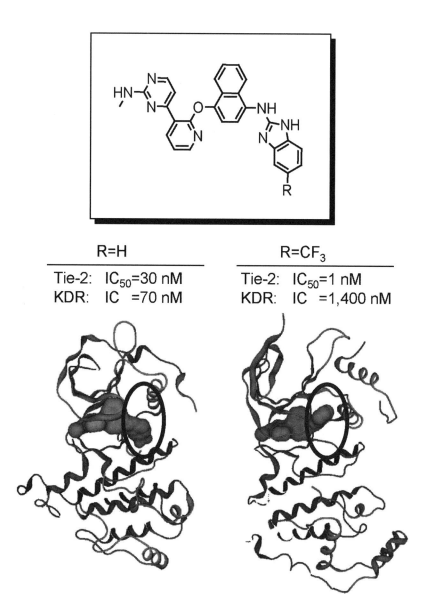

FIGURE 2.1

lectivity, most likely because of a flip in the binding of a terminal benzimidazole group [52]. However, in protein kinases, most, if not all, of the major binding modes have probably been identified and can be used in selectivity analyses [51], although there is a possibility that allosteric sites away from the ATP active site exist [53]. While serendipitous binding modes are an infrequent but important consideration, computational methods are nevertheless useful as systematic, objective analyses for assessing the risk of selectivity issues as well as identifying possible selectivity issues and strategies that should be experimentally considered.

3. DRUGGABILITY

What is "druggability"? It is ultimately the success of the compound in human clinical trials. This includes not only compound properties but also aspects of efficacy, safety, and commercial attractiveness which are difficult to predict. For scientists engaged in drug discovery prior to clinical trials, 'druggability' can be defined more tangibly in terms of the chemical matter at the high-throughput screening and lead generation stages.

3.1 Ligand analysis

Traditionally, druggability has been assessed experimentally. At the HTS stage, teams typically define "druggability" in terms of identifying a "druglike" small molecule with activity in the one micromolar range, where the term "druglike" refers to compounds with physical properties ranges similar to known oral drugs [54–57]. Common "druglike" rules include polar surface area (PSA) less than 140 A^2 [55], number of rotateable bonds less than 10 [56], molecular weight less than about 500 Da, and no more than one rule violation in the Lipinski Rule-of-Five [54]. For a more complete review of "druglike" properties, please see the recent Annual Reports in Computational Chemistry review [58]. Project teams may preferentially identify "leadlike" compounds with lower molecular weights and ClogP values [57]. In the next stage, the lead generation stage, the project team typically defines druggability as the potential to find a compound with nanomolar potency, drug-like properties, as well as experimentally-measured properties related to unwanted secondary pharmacology (for example, selectivity in the CEREP or MDS Panlabs panel), metabolism (microsomal stability, hepatocyte stability, and cytochrome P450 inhibition), and intestinal absorption (permeability, rodent pharmacokinetics).

A group at Abbott has demonstrated that hit-rates from NMR-based fragment screening are a good indicator of the target's druggability [59,60]. The fragment library consisted of "fragmentlike" compounds that have an average molecular weight of 220 and average ClogP of 1.5. Screening a fragment library of around 10,000 compounds using NMR technologies may be more cost effective than screening a full compound library that commonly contain over a million compounds.

3.2 Sequence analysis

In addition to the largely experimental screening approaches, druggability can be assessed based on bioinformatics analysis of the protein sequence. Sequence similarity can be used to determine whether the gene of interest is part of a gene family or sub-family with known druggability status [61]. For instance, aminergic G-protein-coupled receptors (GPCR) and protein kinases are known to be druggable based on marketed drugs as well as collective HTS and medicinal chemistry experience, and so a new aminergic GPCR or protein kinase would be expected to be druggable as well. Hopkins and Groom did a systematic "druggable genome" analysis to identify 130 gene families that are targeted by rule-of-five compliant compounds, and they then identified proteins from the human genome that map to these gene families [61]. The results suggest that only 10% of genes in the human genome map to precedented druggable gene families, and that only 5% are both druggable and disease-relevant. The analysis has been updated by Overington et al. [62] as well as others [63–66], and workers at Novartis have set up a public web server for running a target sequence query against known druggable sequences at http://function.gnf.org/druggable/index.html, although the server is only for academic and non-profit use [66]. Some have pointed out that the much larger "druggable proteome" or "druggable targetome" is more relevant than the "druggable genome" [67]. For instance, the proteasome can be inhibited by a small molecule, and, in addition, there is emerging evidence that protein–protein complexes such as MDM2-p53 are druggable. Nevertheless, the argument that only a small fraction of targets are druggable is not generally contested, and argues for the importance of assessing targets systematically [2,3].

3.3 Structure analysis

Whereas small molecule drugs usually bind to pockets, the reverse is not always true—not all pockets on a biological target are druggable. Upon inspecting crystal structures of druggable and difficult druggability protein binding sites, it becomes clear that druggable pockets tend to be deep, hydrophobic, and of a limited size. Druggable pockets tend to reflect the properties of the drug-like ligands that they bind, and so they might also be called "drug-like binding sites" or "beautiful binding sites" [61,68].

How does one identify beautiful binding sites? Available algorithms include those for identifying ligand-binding "hot spots" on the surface of protein structures, which include fragment-based approaches as well as statistical approaches based on structural descriptors. A computational solvent mapping approach was able to identify known druggable pockets based on known crystal structures, and can further be used to identify hotspots on protein surfaces [69]. Another approach precalculates a van der Waals potential at nodes of a grid that envelops the protein, and then searches for high-scoring grid clusters in order to predict ligand binding pockets [70,71]. Statistical learning approaches include those developed for identifying functional sites [72]. One example is a neural net approach called HotPatch [73] that is based on calculated electrostatic potential, charge, concavity, surface roughness, and hydrophobicity values. HotPatch was successfully used by another group to predict an allosteric small-molecule site in caspases [74]. Another algorithmic approach that is easier to interpret combines probabilistic distribution functions (PDFs) for a similar set of properties [75]. SiteMap (Schrodinger, Inc.) identifies and scores binding sites based on the typical physiochemical properties and, additionally, a "hydrophobic enclosure" term, which accounts for pocket shape in hydrophobic desolvation [76,77].

The approaches discussed so far, however, are focused on predicting pockets for any ligand, as opposed to predicting sites for *druglike* ligands or how druggable a given binding site is. The statistical approaches discussed so far could, in theory, capture druggability given a training set. Work by Hajduk et al. [59] was the first published approach using a statistical method to directly address druggability prediction. They derived a druggability scoring function by performing statistical regression of physiochemical properties calculated for a variety of protein pockets to hit-rates from NMR screening of "leadlike" fragments. More specifically, protein pockets defined using an InsightII (Accelrys, Inc.) flood-fill algorithm were analyzed to generate physiochemical descriptors, including surface area, volume, roughness, and number of charged residues, as well as descriptors of the pocket shape—pocket compactness and three principal moment descriptors. These calculated descriptors were then fitted to NMR fragment-screening data to yield a score, termed the 'druggability index' (D_I):

$$Druggability\ index = -14.0 \cdot X_{PocketCompactness} + 13.6 \cdot \log(X_{PocketCompactness})$$
$$+ 2.98 \cdot \log(X_{ApolarContactArea}) - 0.023 \cdot X_{ApolarContactArea}$$
$$+ 2.98 \cdot \log(X_{SurfaceArea}) - 0.44 \cdot \log(X_{PolarContactArea})$$
$$+ 1.2 \cdot \log(X_{ThirdPrincipalMoment}) - 1.03 \cdot \log(X_{FirstPrincipalMoment})$$
$$+ 0.71 \cdot X_{Roughness}$$
$$- 0.16 \cdot X_{NumberChargedResidues}$$
$$- 1.11.$$

Other descriptors included in the regression (volume, polar surface area, total contact area, and second principal moment) were found to be insignificant and not included in the final equation. With the training set of 23 proteins, the model yielded an r^2 of 0.65 and a q^2 of 0.56. On an external test set of 35 proteins, 94% of the known druggable pockets were correctly predicted as druggable.

In another work, Cheng et al. took an approach that combines a biophysics model with the concept of drug-like physiochemical properties [78]. Intuitively, druggable pockets are hydrophobic [68,79], deep, and have a limited size. The authors used a literature biophysical model for the hydrophobic effect and normalized the equation for drug-like size. The resulting 'Maximal Affinity Prediction' (MAP) equation is an estimate of the maximal affinity of a given binding site for a small molecule with 'druglike' properties.

The MAP score is a continuous score reported as the estimated best K_d achievable by a passively absorbed, non-covalent oral drug:

$$Maximal\ drug\text{-}like\ affinity = -\gamma(r) \cdot \frac{A_{nonpolar}^{target}}{A_{total}^{target}} \cdot 300\ Å^2, \quad where\ \gamma(r) = \frac{45\frac{cal}{mol\cdot Å^2}}{1 - \frac{1.4}{r_{curvature}}}.$$

The surface areas, A, are the measured nonpolar and total surface areas on the defined binding pocket, and, for a concave pocket, the $\gamma(r)$ term represents how easily water will leave a hydrophobic cavity [80–83]. A deeper pocket would have a smaller radius of curvature, $r_{curvature}$, and thus a larger $\gamma(r)$, indicating that water will leave more easily, while a completely flat surface would have an $r_{curvature}$ of infinity. The model is based on a physical model describing hydrophobic free energies of hydrocarbons in water [80,82], from which the authors then normalized the surface area to account for drug-like properties—in particular, they normalized for a druglike molecular weight cut-off of about 550 Da, which is equivalent to about 300 $Å^2$ of surface area [78]. Interestingly, drug-like PSA constraints ($<140\ Å^2$) are accounted for in the model if we assume that the protein pocket PSA complements the ligand PSA. A high polar surface area on the protein pocket will reduce the predicted maximal druglike affinity, and for a pocket with PSA $= 140\ Å^2$ and fairly deep $r_{curvature} = 6\ Å$, the MAP score is 5 µM.

In the MAP model, druggable targets have predicted K_d's in the nM range, while difficult targets had predicted K_d's greater than 100 nM. In the retrospective analysis, several targets were predicted to be difficult targets despite drugs being on the market. Through scholarship they found that these predicted difficult targets were only druggable through a prodrug or active transport approach, pointing out that the approach is useful for predicting passively-absorbed, oral druggability, and other approaches for achieving druggability such as covalent adduct formation, metal chelation, prodrug development, active transport, and allosteric modulation should be kept in mind for difficult targets. In a forward prediction experiment, the authors successfully predicted the druggability of two novel drug targets, fungal homoserine dehydrogenase and haemopoetic prostaglandin D synthase, where druggability was determined by the outcome of a high-throughput screen and subsequent lead optimization at Pfizer [78]. In theory, the lower the maximal affinity of the target binding site, the more freedom the team has in modifying the compound to optimize pharmacodynamic and pharmacokinetic properties while maintaining efficacious potencies. The druggability boundary of 100 nM is generic, and the quantitative predicted K_d values can be useful where a nM affinity inhibitor is, based on physiology, not needed [84].

Although the D_I and MAP equations have different forms, the dominant terms are exceptionally consistent. In the MAP model non-polar surface area and curvature are the properties used, while in the D_I model the highest-weighted descriptors are surface area and pocket compactness (which correlates with curvature).

Since the methods use static crystal structures, one natural issue is how to capture protein flexibility. The MAP approach was robust to differences in the binding site between different co-crystal structures for a set of enzymes where multiple co-crystals were available at the time. The binding sites, however, were all enzyme binding sites that largely consist of stable secondary structure motifs (helices and sheets), and binding sites that are composed of long unstructured regions will certainly see larger variations. For more flexible binding sites, molecular dynamics (MD) simulation could be used. Indeed, applying the D_I approach to snapshots from a MD simulation was necessary and sufficient to correctly predict the druggability of FKBP, Bcl-xL, and AKT-PH domain small-molecule binding sites [85]. These three binding sites involve long loop regions. Another group showed that for three protein-protein binding sites (Bcl-xl, IL-2, and MDM2), MD simulations starting from the apo-crystal conformations successfully resulted in sampling of the known small-molecule bound conformation [86]. Both studies use known druggable proteins, and it would be useful to know if difficult targets can be correctly assessed as well. The lesson here might be that even when starting from ligand-bound structures, care should be taken in assessing druggability of binding sites involving any loop regions, and static structures should not be used at all without simulation of their flexibility if the binding site is formed partially by loop regions. If the experimental conformation in hand is calculated to be sufficiently druggable however, flexibility becomes a non-issue. Flexible protein surfaces can reveal more druggable binding sites than the static structures indicate, as been shown crystallographically in the case of IL-2 [87,88], and calculating the inherent flexibility or adaptability of a site may help in predicting its druggability [89,90].

For targets assessed or found through experience to be difficult, computational methods can aid in lead optimization. In general, structure-based design methods, such as those discussed earlier for identifying hot spots as well as those reviewed in [91], can be useful in driving potency in a more directed manner. Pockets that are difficult to drug tend to be polar, and quantitative charge-optimization approaches can be useful in optimizing leads based on electrostatic interactions, taking into account ligand and receptor desolvation which can be difficult to visualize [92]. Allosteric modulation is increasingly sought [93], and emerging computational methods that combine druggable pocket prediction with functional residue prediction may eventually aid allosteric drug identification [53].

4. CONCLUSIONS

This review covered cheminformatics, bioinformatics, and structure-based drug design approaches and how they aid assessment of selectivity and druggability as well as setting of lead optimization strategies. While computational approaches will continue to improve in accuracy, they are nevertheless useful today for bringing together data in a rational, model-based manner to inform experiments and decision making.

REFERENCES

1. Brown, D., Superti-Furga, G. Rediscovering the sweet spot in drug discovery. Drug Discov. Today 2003, 8, 1067–77.

2. Stahl, M., Guba, W., Kansy, M. Integrating molecular design resources within modern drug discovery research: The Roche experience. Drug Discov. Today 2006, 11, 326–33.

3. Frearson, J.A., Wyatta, P.G., Gilberta, I.H., Fairlamba, A.H. Target assessment for antiparasitic drug discovery. Trends Parasitol. 2007, 23, 589–95.

4. Loging, W., Harland, L., Williams-Jones, B. High-throughput electronic biology: Mining information for drug discovery. Nature Rev. Drug. Discov. 2007, 6, 220–30.

5. Daub, H., Specht, K., Ullrich, A. Strategies to overcome resistance to targeted protein kinase inhibitors. Nature Rev. Drug Disc. 2004, 3, 1001–10.

6. Knight, Z.A., Shokat, K.M. Features of selective kinase inhibitors. Chem. Biol. 2005, 12, 621–37.

7. Rockey, W.M., Elcock, A.H. Rapid computational identification of the targets of protein kinase inhibitors. J. Med. Chem. 2005, 48, 4138–52.

8. Vieth, M., Higgs, R.E., Robertson, D.H., Shapiro, M., Gragg, E.A., Hemmerle, H. Kinomics-structural biology and chemogenomics of kinase inhibitors and targets. Biochim. Biophys. Acta 2004, 1697, 243–57.

9. Vieth, M., Sutherland, J.J., Robertson, D.H., Campbell, R.M. Kinomics: Characterizing the therapeutically validated kinase space. Drug Discov. Today 2005, 10, 839–46.

10. Fox, T., Kriegl, J.M. Linear quantitative structure–activity relationships for the interaction of small molecules with human cytochrome P450 isoenzymes. Ann. Reports Comp. Chem. 2005, 1, 63–81.

11. Verras, A., Kuntz, I.D., Ortiz de Montellano, P.R. Cytochrome P450 enzymes: Computational approaches to substrate prediction. Ann. Reports Comp. Chem. 2005, 1, 171–95.

12. Clark, D.E. Computational prediction of ADMET properties: Recent developments and future challenges. Ann. Reports Comp. Chem. 2005, 1, 133–51.

13. Oprea, T.I., Tropsha, A. Target, chemical and bioactivity databases—Integration is key. Drug Disc. Today: Technologies 2006, 3, 357–66.

14. Nidhi, Glick, M., Davies, J.W., Jenkins, J.L. Prediction of biological targets for compounds using multiple-category Bayesian models trained on chemogenomics databases. J. Chem. Inf. Model. 2006, 46, 1124–33.

15. Frye, S.V. Structure-activity relationship homology (SARAH): A conceptual framework for drug discovery in the genomic era. Chem Biol. 1999, 6, R3–7.

16. Greenbaum, D.C., Arnold, W.D., Lu, F., Hayrapetian, L., Baruch, A., Krumrine, J., Toba, S., Chehade, K., Brömme, D., Kuntz, I.D., Bogyo, M. Small molecule affinity fingerprinting. A tool for enzyme family subclassification, target identification, and inhibitor design. Chem. Biol. 2002, 9, 1085–94.

17. Martin, Y.C., Kofron, J.L., Traphagen, L.M. Do structurally similar molecules have similar biological activity? J. Med. Chem. 2002, 45, 4350–8.

18. Paolini, G.V., Shapland, R.H.B., van Hoorn, W.P., Mason, J.S., Hopkins, A.L. Global mapping of pharmacological space. Nature Biotech. 2006, 24, 805–15.

19. Keiser, M.J., Roth, B.L., Armbruster, B.N., Ernsberger, P., Irwin, J.J., Shoichet, B.K. Relating protein pharmacology by ligand chemistry. Nature Biotech. 2007, 25, 197–206.

20. Altschul, S.F., Gish, W., Miller, W., Myers, E.W., Lipman, D.J. Basic local alignment search tool. J. Mol. Biol. 1990, 215, 403–10.

21. Hert, J., Keiser, M.J., Irwin, J.J., Oprea, T.I., Shoichet, B.K. Quantifying the relationships among drug classes. J. Chem. Inf. Model. 2008, 48, 755–65.

22. Nidhi, Glick, M., Davies, J.W., Jenkins, J.L. Prediction of biological targets for compounds using multiple-category Bayesian models trained on chemogenomics databases. J. Chem. Inf. Model. 2006, 46, 1124–33.

23. Mestres, J., Martín-Couce, L., Gregori-Puigjané, E., Cases, M., Boyer, S. Ligand-based approach to in silico pharmacology: Nuclear receptor profiling. J. Chem. Inf. Model. 2006, 46, 2725–36.

24. Izrailev, S., Farnum, M.A. Enzyme classification by ligand binding. Proteins 2004, 57, 711–24.

25. Pruitt, K.D., Tatusova, T., Maglott, D.R. NCBI Reference Sequence (RefSeq): A curated non-redundant sequence database of genomes, transcripts and proteins. Nucleic Acids Res. 2007, 35, D61–5.

26. Higgins, D.G., Sharp, P.M. CLUSTAL: A package for performing multiple sequence alignment on a microcomputer. Gene 1988, 73, 237–44.

27. Caffrey, D.R., Dana, P.H., Mathur, V., Ocano, M., Hong, E.J., Wang, Y.E., Somaroo, S., Caffrey, B.E., Potluri, S., Huang, E.S. PFAAT version 2.0: A tool for editing, annotating, and analyzing multiple sequence alignments. BMC Bioinformatics 2007, 8, 381.
28. Clamp, M., Cuff, J., Searle, S.M., Barton, G.J. The jalview java alignment editor. Bioinformatics 2004, 20, 426–7.
29. Tamura, K., Dudley, J., Nei, M., Kumar, S. MEGA4: Molecular Evolutionary Genetics Analysis (MEGA) software version 4.0. Mol. Biol. Evolution 2007, 24, 1596–9.
30. Lee, M.R., Dominguez, C. MAP kinase p38 inhibitors: Clinical results and an intimate look at their interactions with p38alpha protein. Curr. Med. Chem. 2005, 12, 2979–94.
31. Vulpetti, A., Bosotti, R. Sequence and structural analysis of kinase ATP pocket residues. Farmaco. 2004, 59, 759–65.
32. Kothe, M., Kohls, D., Low, S., Coli, R., Rennie, G.R., Feru, F., Kuhn, C., Ding, Y.H. Selectivity-determining residues in Plk1. Chem. Biol. Drug Des. 2007, 70, 540–6.
33. Kothe, M., Kohls, D., Low, S., Coli, R., Cheng, A.C., Jacques, S.L., Johnson, T.L., Lewis, C., Loh, C., Nonomiya, J., Sheils, A.L., Verdries, K.A., Wynn, T.A., Kuhn, C., Ding, Y.H. Structure of the catalytic domain of human polo-like kinase 1. Biochemistry 2007, 46, 5960–71.
34. Sheinerman, F.B., Giraud, E., Laoui, A. High affinity targets of protein kinase inhibitors have similar residues at the positions energetically important for binding. J. Mol. Biol. 2005, 352, 1134–56.
35. Ortiz, A.R., Gomez-Puertas, P., Leo-Macias, A., Lopez-Romero, P., Lopez-Viñas, E., Morreale, A., Murcia, M., Wang, K. Computational approaches to model ligand selectivity in drug design. Curr. Top. Med. Chem. 2006, 6, 41–55.
36. Lichtarge, O., Bourne, H.R., Cohen, F.E. An evolutionary trace method defines binding surfaces common to protein families. J. Mol. Biol. 1996, 257, 342–58.
37. Armon, A., Graur, D., Ben-Tal, N. ConSurf: An algorithmic tool for the identification of functional regions in proteins by surface mapping of phylogenetic information. J. Mol. Biol. 2001, 307, 447–63.
38. Schmitt, S., Kuhn, D., Klebe, G. A new method to detect related function among proteins independent of sequence and fold homology. J. Mol Biol. 2002, 323, 387–406.
39. Ferre, F., Ausiello, G., Zanzoni, A., Helmer-Citterich, M. Functional annotation by identification of local surface similarities: A novel tool for structural genomics. BMC Bioinformatics 2005, 6, 194.
40. Gold, N.D., Jackson, R.M. Fold independent structural comparisons of protein–ligand binding sites for exploring functional relationships. J. Mol. Biol. 2006, 355, 1112–24.
41. Kuhn, D., Weskamp, N., Hüllermeier, E., Klebe, G. Functional classification of protein kinase binding sites using Cavbase. Chem. Med. Chem. 2007, 2, 1432–47.
42. Gold, N.D., Deville, K., Jackson, R.M. New opportunities for protease ligand-binding site comparisons using SitesBase. Biochem. Soc. Trans. 2007, 35, 561–5.
43. Pastor, M., Cruciani, G. A novel strategy for improving ligand selectivity in receptor-based drug design. J. Med. Chem. 1995, 38, 4637–47.
44. Kastenholz, M.A., Pastor, M., Cruciani, G., Haaksma, E.E.J., Fox, T. GRID/CPCA: A new computational tool to design selective ligands. J. Med. Chem. 2000, 43, 3033–44.
45. Myshkin, E., Wang, B. Chemometrical classification of ephrin ligands and Eph kinases using GRID/CPCA approach. J. Chem. Inf. Comput. Sci. 2003, 43, 1004–10.
46. Vulpetti, A., Crivori, P., Cameron, A., Bertrand, J., Brasca, M.G., D'Alessio, R., Pevarello, P. Structure-based approaches to improve selectivity: CDK2-GSK3beta binding site analysis. J. Chem. Inf. Model. 2005, 45, 1282–90.
47. Yoon, S., Smellie, A., Hartsough, D., Filikov, A. Computational identification of proteins for selectivity assays. Proteins. 2005, 59, 434–43.
48. Sherman, W., Tidor, B. Novel method for probing the specificity binding profile of ligands: Applications to HIV protease. Chem. Biol. Drug Des. 2008, 71, 387–407.
49. Radhakrishnan, M.L., Tidor, B. Specificity in molecular design: A physical framework for probing the determinants of binding specificity and promiscuity in a biological environment. J. Phys. Chem. B 2007, 111, 13419–35.
50. Watkins, R.E., Wisely, G.B., Moore, L.B., Collins, J.L., Lambert, M.H., Williams, S.P., Willson, T.M., Kliewer, S.A., Redinbo, M.R. The human nuclear xenobiotic receptor PXR: Structural determinants of directed promiscuity. Science 2001, 292, 2329–33.

51. Jacobs, M.D., Caron, P.R., Hare, B.J. Classifying protein kinase structures guides use of ligand-selectivity profiles to predict inactive conformations: Structure of lck/imatinib complex. Proteins 2007, 70, 1451–60.
52. Cee, V.J., Cheng, A.C., Romero, K., Bellon, S., Mohr, C., Whittington, D.A., Bready, J., Caenepeel, S., Coxon, A., Deak, H.L., Hodous, B.L., Kim, J.L., Lin, J., Nguyen, H., Olivieri, P.R., Patel, V.F., Wang, L., Hughes, P., Geuns-Meyer, S., Pyridyl-pyrimidine benzimidazole derivatives as potent, selective, and orally bioavailable inhibitors of Tie-2 kinase, Bioorg. Med. Chem. Ltrs., in press.
53. Coleman, R.G., Salzberg, A.C., Cheng, A.C. Structure-based identification of small molecule binding sites using a free energy model. J. Chem. Inf. Model. 2006, 46, 2631–7.
54. Lipinski, C.A., Lombardo, F., Dominy, B.W., Feeney, P.J. Experimental and computational approaches to estimate solubility and permeability in drug discovery and development settings. Adv. Drug Delivery Rev. 2001, 46, 3–26.
55. Palm, K., Stenberg, P., Luthman, K., Artursson, P. Polar molecular surface properties predict the intestinal absorption of drugs in humans. Pharm. Res. 1997, 14, 568–71.
56. Veber, D.F., Johnson, S.R., Cheng, H.Y., Smith, B.R., Ward, K.W., Kopple, K.D. Molecular properties that influence the oral bioavailability of drug candidates. J. Med. Chem. 2002, 45, 2615–23.
57. Oprea, T.I., Davis, A.M., Teague, S.J., Leeson, P.D. Is there a difference between leads and drugs? A historical perspective. J. Chem. Inf. Comput. Sci. 2001, 41, 1308–15.
58. Lipinski, C. Filtering in drug discovery. Ann. Reports Comput. Chem. 2005, 1, 155–68.
59. Hajduk, P.J., Huth, J.R., Fesik, S.W. Druggability indices for protein targets derived from NMR-based screening data. J. Med. Chem. 2005, 48, 2518–25.
60. Hajduk, P.J., Huth, J.R., Tse, C. Predicting protein druggability. Drug Discov. Today 2005, 10, 1675–82.
61. Hopkins, A.L., Groom, C.R. The druggable genome. Nat. Rev. Drug Discov. 2002, 1, 727–30.
62. Overington, J.P., Al-Lazikani, B., Hopkins, A.L. How many drug targets are there? Nature Rev. Drug Discov. 2006, 5, 993–6.
63. Sakharkar, M.K., Sakharkar, K.R., Pervaiz, S. Druggability of human disease genes. Int. J. Biochem. Cell Biol. 2007, 39, 1156–64.
64. Hambly, K., Danzer, J., Muskal, S., Debe, D.A. Interrogating the druggable genome with structural informatics. Mol. Divers. 2006, 10, 273–81.
65. Russ, A.P., Lampel, S. The druggable genome: An update. Drug Discov. Today 2005, 10, 1607–10.
66. Orth, A.P., Batalov, S., Perrone, M., Chanda, S.K. The promise of genomics to identify novel therapeutic targets. Expert Opin. Ther. Targets 2004, 8, 587–96.
67. Kubinyi, H. Drug research: Myths, hype and reality. Nature Rev. Drug Disc. 2003, 2, 665–8.
68. Fauman, E.B., Hopkins, A.L., Groom, C.R. Structural bioinformatics in drug discovery. In: Weissig, H., Bourne, P., editors. Structural Bioinformatics. Hoboken, NJ: Wiley-Liss; 2003, p. 477–98.
69. Landon, M.R., Lancia Jr., D.R., Yu, J., Thiel, S.C., Vajda, S. Identification of hot spots within druggable binding regions by computational solvent mapping of proteins. J. Med. Chem. 2007, 50, 1231–40.
70. An, J., Totrov, M., Abagyan, R. Pocketome via comprehensive identification and classification of ligand binding envelopes. Mol. Cell Proteomics. 2005, 4, 752–61.
71. An, J., Totrov, M., Abagyan, R. Comprehensive identification of druggable protein ligand binding sites. Genome Inform. 2004, 15, 31–41.
72. Nayal, M., Honig, B. On the nature of cavities on protein surfaces: Application to the identification of drug-binding sites. Proteins 2006, 63, 892–906.
73. Pettit, F.K., Bare, E., Tsai, A., Bowie, J.U. HotPatch: A statistical approach to finding biologically relevant features on protein surfaces. J. Mol. Biol. 2007, 369, 863–79.
74. Hardy, J.A., Lam, J., Nguyen, J.T., O'Brien, T., Wells, J.A. Discovery of an allosteric site in the caspases. Proc. Natl. Acad. Sci. USA 2004, 101, 12461–6.
75. Joughin, B.A., Tidor, B., Yaffe, M.B. A computational method for the analysis and prediction of protein: Phosphopeptide-binding sites. Protein Sci. 2005, 14, 131–9.
76. Halgren, T. New method for fast and accurate binding-site identification and analysis. Chem. Biol. Drug Des. 2007, 69, 146–8.
77. Friesner, R.A., Murphy, R.B., Repasky, M.P., Frye, L.L., Greenwood, J.R., Halgren, T.A., Sanschagrin, P.C., Mainz, D.T. Extra precision Glide: Docking and scoring incorporating a model of hydrophobic enclosure for protein–ligand complexes. J. Med. Chem 2006, 49, 6177–96.

78. Cheng, A.C., Coleman, R.G., Smyth, K.T., Cao, Q., Soulard, P., Caffrey, D.R., Salzberg, A.C., Huang, E.S. Structure-based maximal affinity model predicts small-molecule druggability. Nature Biotech. 2007, 25, 71–5.

79. Davis, A.M., Teague, S.J. Hydrogen bonding, hydrophobic interactions, and failure of the rigid receptor hypothesis. Agnew. Chem. Int. Ed. 1999, 38, 736–49.

80. Sharp, K.A., Nicholls, A., Fine, R.F., Honig, B. Reconciling the magnitude of the microscopic and macroscopic hydrophobic effects. Science 1991, 252, 106–9.

81. Cheng, Y.-K., Rossky, P.J. Surface topography dependence of biomolecular hydrophobic hydration. Nature 1998, 392, 696–9.

82. Southall, N.T., Dill, K.A. The mechanism of hydrophobic solvation depends on solute radius. J. Phys. Chem. B 2000, 104, 1326–31.

83. De Young, L.R., Dill, K.A. Partitioning of nonpolar solutes into bilayers and amorphous *n*-alkanes. J. Phys. Chem. 1990, 94, 801–9.

84. Copeland, R.A., Pompliano, D.L., Meek, T.D. Drug-target residence time and its implications for lead optimization. Nature Rev. Drug Discov. 2006, 5, 730–9.

85. Brown, S.P., Hajduk, P.J. Effects of conformational dynamics on predicted protein druggability. Chem. Med. Chem. 2006, 1, 70–2.

86. Eyrisch, S., Helms, V. Transient pockets on protein surfaces involved in protein–protein interaction. J. Med. Chem. 2007, 50, 3457–64.

87. Braisted, A.C., Oslob, J.D., DeLano, W.L., Hyde, J., McDowell, R.S., Waal, N., Yu, C., Arkin, M.R., Raimundo, B.C. Discovery of a potent small molecule IL-2 inhibitor through fragment assembly. J. Am. Chem. Soc. 2003, 125, 3714–5.

88. Raimundo, B.C., Oslob, J.D., Braisted, A.C., Hyde, J., McDowell, R.S., Randal, M., Waal, N.D., Wilkinson, J., Yu, C.H., Arkin, M.R. Integrating fragment assembly and biophysical methods in the chemical advancement of small-molecule antagonists of IL-2: An approach for inhibiting protein-protein interactions. J. Med. Chem. 2004, 47, 3111–30.

89. Thanos, C.D., Randal, M., Wells, J.A. Potent small-molecule binding to a dynamic hot spot on IL-2. J. Am. Chem. Soc. 2003, 125, 15280–1.

90. Thanos, C.D., DeLano, W.L., Wells, J.A. Hot-spot mimicry of a cytokine receptor by a small molecule. Proc. Natl. Acad. Sci. USA 2006, 103, 15422–7.

91. Joseph-McCarthy, D. Structure-based lead optimization. Ann. Reports Comp. Chem. 2005, 1, 169–83.

92. Armstrong, K.A., Tidor, B., Cheng, A.C. Optimal charges in lead progression: A structure-based neuraminidase case study. J. Med. Chem. 2006, 49, 2470–7.

93. Hardy, J.A., Wells, J.A. Searching for new allosteric sites in enzymes. Curr. Opin. Struct. Biol. 2004, 14, 706–15.

Section 2
Biological Modeling

Section Editor: Nathan Baker

Department of Biochemistry and Molecular Biophysics
Washington University
700 S. Euclid Ave.
Room 113, Campus Box 8036
St. Louis, MO 63110
USA

CHAPTER 3

Machine Learning for Protein Structure and Function Prediction

Robert Ezra Langlois* and **Hui Lu*,[1]**

Contents

1. INTRODUCTION

Machine learning is an established tool in many problem domains ranging from computer vision to stock markets to computational chemistry. A machine learning algorithm automatically discovers patterns in historical data to improve future

* Bioinformatics Program, Department of Bioengineering, University of Illinois-Chicago, 851 S Morgan, Room 218, Chicago, IL 60607, USA. E-mails: ezra@uic.edu (R.E. Langlois), huilu@uic.edu (H. Lu).
[1] Corresponding author.

Annual Reports in Computational Chemistry, Vol. 4
ISSN 1574-1400, DOI: 10.1016/S1574-1400(08)00003-0

decisions or actions in complex applications. In biological and medical applications [1,2], often having large amounts of data, machine learning has shown great promise in replacing some "wet-lab" experiments, guiding research, and elucidating underlying interactions within the data. For example, a learner can be trained to identify functional residues on a protein, which may, in turn, suggest potential binding mechanisms; such mechanisms can be further validated by mutagenesis experiments. While a prediction can suggest a potential function of a protein, the rules learned to make this prediction can provide further insight into underlying mechanisms governing the functional interaction. In other words, these rules characterize underlying mechanisms or interactions in the form of features [3] and their relationships.

Machine learning is primarily concerned with developing algorithms that "learn" and has deep roots in both artificial intelligence and statistics. Recently, due to the ever increasing amount of available computer power and data storage, machine learning has increased proportionally in popularity and has become essential in many fields. In particular, neural networks and decision trees represent two algorithms in machine learning that have been in main stream use for many years. Many state-of-the-art learning algorithms have been developed on the basis of these two algorithms. Indeed, neural networks [4] have given rise support vector networks [5] (support vector machines, SVM), Bayesian networks [6], conditional Markov random fields [7], among others. Likewise, decision trees [8] form the basis of many methods such as boosting [9], bagging [10], random forests [11], among others. There exist two fundamental types of machine learning algorithms derived from differing views in statistics: frequentist and evidential [12]. Decision trees and neural networks belong to the former whereas Bayesian networks and conditional Markov random fields belong to the latter.

Machine learning has gained popularity in biology with the analysis of high throughput experiments such as microarray data analysis [13,14]. In recent years, the machine learning approach has been extensively used in protein structure and function modeling, which is the current focus of this review. While the application of machine learning to protein structure and function modeling is no more difficult than to microarray data analysis, its later adoption is due partly to a greater focus on biophysical approaches and previously limited number of examples.

The outline of this chapter is as follows. The second section introduces a number of important supervised learning problems and illustrates how a biomolecular application can be cast in each problem formulation. Specifically, modeling protein–DNA interactions serves as the example for each of these formulations. The third section summarizes recent applications of machine learning to biomolecular modeling. The final section discusses current trends and future directions of machine learning applications to biomolecular modeling.

2. MACHINE LEARNING PROBLEM FORMULATIONS

Machine learning can be broken down into a number of problem formulations. The three major categories comprise supervised, unsupervised and reinforcement learning. In supervised learning, the algorithm is given a set of input and output vectors in order to learn a mapping between input and output. In reinforcement learning, the algorithm searches for appropriate actions for a given situation in order to maximize some reward; in other words, the learner is not given optimal outputs as in supervised learning but must learn the outputs under some guidance. In unsupervised learning, the algorithm is only given input vectors and must find some internal representation of the input data either by clustering similar objects into groups or estimating the distribution (density) of the data. Each of these learning formulations has found applications in biology. For instance, both supervised and unsupervised learning have been used in the analysis of microarrays [13,14]. Likewise, reinforcement learning has been applied to the identification of unique protein fragments [15], which have been applied to *ab initio* prediction [16] and homology detection of proteins [17]. Here, we will focus on supervised learning problems and for clarity each formulism will be accompanied with an example that involves modeling protein–DNA interactions. Most examples are taken from published work and ongoing research in the lab.

Before continuing the discussion of machine learning problems, let us first review the importance of protein–DNA interactions. In a cell, proteins interact with DNA to replicate, repair and regulate DNA-centric processes; we refer to these proteins as DNA-binding proteins (Figure 3.1). They comprise roughly 7% of proteins encoded in the eukaryotic genome and 6% in the prokaryotic genome [19]. They also represent a diverse set sequences, structures and functions. For example, Luscombe et al. [20] classified DNA-binding proteins into 54 structural families. A DNA-binding protein uses specific site(s) (set of residues) to bind to a DNA sequence; these specific interactions have received commiserate interest from molecular biologists who have developed a number of techniques to investi-

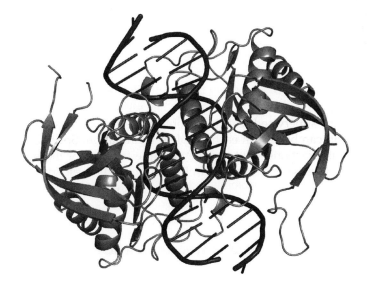

FIGURE 3.1   A restriction endonuclease BglII bound to DNA (1DFM) rendered in PyMOL [18].

Table 3.1 Supervised machine learning problems

Problem	Input	Label	Loss
Binary[a]	$(\vec{x}, y)$	$y \in \{+1, -1\}$	$\Pr(h(\vec{x}) \neq y)$
Multi-class[a]	$(\vec{x}, y)$	$y \in \{0, 1, \ldots, m\}$	$\Pr(h(\vec{x}) \neq y)$
Ranking	$(\vec{x}, y)$	$y \in \{+1, -1\}$	$E[\frac{\sum_{i \neq j} 1(y_i > y_j) \pi(x_i, x_j)}{\sum_{i<j} 1(y_i > y_j)}]$
Regression	$(\vec{x}, y)$	$y \in \mathbb{R}$	$E[(h(\vec{x}) - y)^2]$
Importance[b]	$(\vec{x}, y, w)$	$y \in \{+1, -1\}$	$E[wI(h(\vec{x}) \neq y)]$
MIL[c]	$(\vec{x}_1, \ldots, \vec{x}_{n(i)}, y)$	$y \in \{+1, -1\}$	$\Pr(h(\mathbf{B}) \neq y)$
Structured[d]	$(\vec{x}, y, \vec{c})$	$y \in \mathbb{Y}$	$E[c_{h(x)}]$

a Classification.

b Importance-weighted classification.

c Multiple-instance learning.

d Structured-prediction.

gate mechanisms governing protein–DNA interactions [21–25]. A computational approach to identify these sites not only aids in such an investigation but also serves to guide future experiments.

For the purpose of the discussion on machine learning, we will use the following definitions and notation. Let $(\vec{x}, y)$ be a labeled example where $x_j \in \vec{x}$, $j = 1, \ldots, m$ is a vector of numerical attributes x_j and y is an associated label. Let m be fixed such that every attribute vector has the same length. We denote the probability of an event as Pr and the probabilistic expectation of an event as E. In supervised learning, we are given a set of examples $(\vec{x}_i, y_i)$, $i = 1, \ldots, n$ called *training data*. From this data, a learning algorithm L attempts to find a hypothesis h minimizing some loss l and maps $\vec{x}$ to y, i.e. $\hat{y} \leftarrow h(\vec{x})$. The *loss* measures how well the hypothesis maps the input training vectors to the corresponding output class label; in other words, how many mistakes the hypothesis makes when predicting the label of an input vector. This hypothesis can then be evaluated on its ability to generalize the training data to unseen *testing data*. A supervised learning problem largely depends on three elements: input vector data $(\vec{x})$, output class (y) and loss (l) (see Table 3.1). Consider DNA-binding protein prediction, x can be a 20-dimensional vector describing the composition of amino acids in a protein and y can be the either binding to DNA or not. The training data comprises the known DNA binding and non-binding proteins and the loss function can be the percentage of misclassified proteins.

2.1 Classification

Classification is probably the most common supervised machine learning formalism where an example is assigned a grouping based on some hypothesis learned over a set of training examples $(\vec{x}, y)$. A classification algorithm (or classifier) searches for hypothesis h that minimizes the classification error $e =$

$Pr(h(\vec{x}) \neq y)$. One common classification problem is binary classification where an example is placed in one of two mutually exclusive groups, $y \in \{1, -1\}$. Note that many binary classifiers produce a real-valued output such that $\mathbb{R} \leftarrow h(\vec{x})$. A threshold is applied to the real value, $\text{sign}_t[h(\vec{x})] \in \{1, -1\}$ such that if the real value exceeds the threshold t then the output is 1 otherwise the output is -1.

The performance of a classifier may be measured by a number of metrics. We have four basic counts to tabulate a binary prediction: TP (True Positive), FP (False Positive), TN (True Negative), and FN (False Negative). Most metrics are calculated from these four numbers. A standard classifier minimizes the error estimated by the number of mistakes over the number of predictions; this is often measured by the accuracy $(TP + TN)/(TP + TN + FP + FN)$. Nevertheless, a binary classifier can make two types of errors, one for each class. For the positive class it is called sensitivity $TP/(TP + FP)$. Similarly, the for the negative class it is called specificity: $TN/(TN + FN)$. Note that each of these metrics depends on the threshold used for a real-valued classifier, e.g. a higher threshold will lower the sensitivity and increase the specificity.

An example binary classification task is to predict whether a given protein binds DNA using sequence- and structure-based information. In this task, the data set comprises a set of non-homologous proteins where the positive examples bind DNA and the negatives do not. Each protein can be represented as a set of features describing sequence- and structure-based characteristics. Typical features used in protein representation include sequence composition, hydrophobic patterns, evolutionary conservation, among others. For the specific nature of protein–DNA interaction, the features related to electrostatics such as charge of the protein and the surface positive electrostatic patch play a dominant role in distinguishing the binding behavior [26]. DNA binding protein prediction has been tackled by a number of published works using classifiers such as hidden Markov models [27,28], neural networks [29,30], support vector machines [3,26,31], logistic regression [32] and boosted trees [3]. In terms of classification, the best published results are about 86% [26] to 88% [3] accuracy.

A number of works have also explored using only primary sequence to represent a protein, often using a set of physio-chemical properties [33–35]. Indeed, the importance of a machine learning approach to sequence function assignment lies, in that, it works irrespective of sequence similarity. To this end, the data sets in published work are usually non-redundant: no two sequences share more than a certain percentage of sequence identity i.e. 40% [35], 25% [33] and 20% [34]. These approaches achieved 86%, 70% and 77% (RNA/DNA together) accuracy in distinguishing DNA-binding proteins from non-binding proteins, respectively. Note that for each work, the data set is significantly different and the only direct comparison was carried out between Fujishima et al. [34] and Cai et al. [35] where the difference in terms of accuracy was about 1%.

Another example binary classification task is to predict whether a given residue in the protein binds DNA. The data set for this task comprises a set of non-homologous structures solved in complex with DNA, which are decomposed into individual residues. The positive examples are surface residues close to DNA and all other surface residues are negative. Each residue can be represented as a set of features comprising residue type, secondary structure or predicted secondary structures, composition of residues within a predetermined radius, etc. This problem has also been investigated by a number of published works using classifiers including neural networks [36] support vector machines [31,37,38] and Naïve Bayes [39]. The performance of these methods ranges from 70% for sequence alone [38] to 78% using sequence plus evolution information [39] to 82% using sequence and structure and evolution [37].

The problem of classification can be extended to multiple classes where an example is assigned to one of several mutually exclusive classes, $y \in \{0, 1, \ldots, k\}$ where $k > 2$. An example multi-class classification task is to predict the type of interaction between residue and DNA, which can be broken down into three classes: side-chain sugar, side-chain base and none. The training data remains the same as the previous example except the label on each example is now one of these three classes: side-chain sugar, side-chain base and none. Another example can be the prediction of proteins with properties of DNA binding, RNA binding, and no binding to nucleic acid.

2.2 Ranking

Ranking is a supervised learning technique that attempts to order predictions such that the top ranked predictions are more likely to be the class of interest. Since many classifiers produce a confidence in prediction (e.g. for support vector machines this is the distance from the margin), these same classifiers can be treated as ranking algorithms. Thus, the setup of this problem is very similar to classification except instead of measuring the results in terms of error, a ranking metric such as the area under the receiver operating characteristic curve (AUR) is used. The receiver operating characteristic curve plots the sensitivity versus the one minus the specificity over all thresholds in a real-valued binary classification system. A standard ranking algorithm attempts to minimize the expected AUR as follows:

$$l = E\left[\frac{\sum_{i \neq j} I(y_i > y_j)\pi(x_i, x_j)}{\sum_{i < j} 1(y_i > y_j)}\right]$$

where $I(\cdot)$ gives one if the expression is true and zero otherwise and $\pi(x_i, x_j)$ gives one if x_i, x_j are in the correct order and zero otherwise [40].

An example ranking problem is to predict the top n residues most likely to bind DNA. This is motivated by the problem where an experimentalist wishes to make several mutations in a protein to find the most important binding site and could use guidance from a machine learning algorithm. Likewise, one may wish submit multiple proteins to a server and retrieve a ranking of the top n most likely to bind DNA. Most classifiers can serve as ranking algorithms (although they do not minimize the AUR) such that the method of evaluation determines which problem is solved. To this end, many publications analyze the AUR or the receiver operating characteristic plot in order to measure the ability of their technique to rank more relevant examples (e.g. proteins that bind DNA) higher than irrelevant ones [3, 32]. In terms of ranking, the best published results range from 88% [3] to 93% [32]

area under the receiving operator curve (on two different data sets) when ranking potential DNA-binding proteins over non-binding proteins.

2.3 Regression

Regression is a supervised learning technique that attempts to assign a real-valued output to an example. Similar to classification, a regressor learns a hypothesis over a set of training examples $(\vec{x}, y)$; however, in regression, the class value is some real number, $y \in \mathbb{R}$. A regression algorithm searches for hypothesis $\mathbb{R} \leftarrow h(\vec{x})$ that minimizes the mean squared error $e = \mathrm{E}[(h(\vec{x}) - y)^2]$. Note that this formulation has been widely used in drug design with QSAR [41], which is short for quantitative structure-activity relationship.

An example regression task is to predict the affinity of a protein binding to DNA. There are a number of experimental methods including ChiP-chip [22,23] and footprinting [21] to ascertain protein–DNA binding affinity. Indeed, an efficient approach to predicting binding affinity involves training a regressor over example DNA-binding sites where the label corresponds to their affinity for DNA derived from ChiP-chip data [42]. The results of published work demonstrate that this method better predicts affinities in 86% of the cases when compared to PSSM (position-specific position matrix).

2.4 Importance-weighted classification

Importance-weighted classification assigns a higher cost to misclassifying specific examples over others. In this problem, every training example is given a weight, $(\vec{x}, y, w)$. The resulting classifier searches for hypothesis h that minimizes the weighted misclassification error $e = \mathrm{E}[wI(h(\vec{x}) \neq y)]$ where $I(\cdot)$ gives one if the expression is true and zero otherwise. The output of the hypothesis remains $h(\vec{x}) \in \{1, -1\}$ and the test cases are unweighted, $(\vec{x}, y)$. Note that a special case of importance-weighted classification is cost-sensitive classification where each class is given a weight i.e. the weight on each example depends on its class.

Generally, a residue is classified as DNA-binding if it is found within a specific radius of DNA. However, under this definition, a residue found in the vicinity of DNA regardless of actual interaction is assigned as DNA-binding. Assigning a weight to training examples based on interaction type and/or count biases the learning algorithm toward residues more likely to bind based on information known *a priori*. Likewise, given that there are considerably less binding residues compared to non-binding, a cost-sensitive classification problem (special case of importance-weighted) can be used to maintain a balanced accuracy in terms of sensitivity and specificity. When balancing sensitivity and specificity of a neural network, published work has achieved a balanced accuracy of 64% [43].

2.5 Multiple-instance learning

In the multiple-instance learning (MIL) problem, examples are organized into groups called bags and the label is associated with the bag, not the example. A bag is labeled positive if at least one example in the bag is positive, otherwise the bag is negative. Note that it is unknown which example in the positive bag is positive only that at least one example in the positive bag is positive. Formally, unlabeled examples $\vec{x}$ are organized into bags $\vec{x} \in \mathbf{B}$ and each bag has an associated label $y \in \{0, 1\}$. The training data for this problem comprises labeled sets of bags $\{\mathbf{B}, y\}$. An MIL algorithm searches for hypothesis $h(\mathbf{B}) \in \{0, 1\}$ that minimizes the bag-level classification error $e = \Pr(h(\mathbf{B}) \neq y)$ where $h(\mathbf{B}) = h(\vec{x}_1) \vee h(\vec{x}_2) \vee \cdots \vee h(\vec{x}_m)$ and $h(\vec{x}_1) \in \{0, 1\}$. The symbol $\vee$ denotes the logical OR operator. Note that multiple-instance learning can be seen as classification with positive class noise.

Consider the residue–DNA interaction problem where the only available training data consists of a set of non-homologous proteins consisting of ones known to bind and not to bind DNA. The interacting residues are not known. In the MIL problem formulation, the residues form examples and the proteins bags. Note that a protein, which binds DNA, must have at least one residue (usually more) that binds DNA and a non-binding protein will not have any such residues; this satisfies the conditions of MIL. Based on this data, an MIL algorithm can learn to predict which residues bind DNA without having labeled residues in the training data. This formalism is very attractive since there are many proteins with known function lacking the information of specific functional residues. Applying MIL in DNA binding residue prediction can achieve similar accuracy when compared with binary classification, but requires only the bag level information (unpublished work).

2.6 Structured-prediction

In the classification setting, every example is assumed to be independent of every other example. However, there are cases where dependencies exist between examples and the label corresponding to such dependent examples forms a complex object. The structured-prediction problem attempts to predict a complex object such as a protein interaction network, phylogenetic tree, a binding site on a protein, etc. In short, a structured-prediction problem D is a classification problem where $y \in \mathbb{Y}$ (the space of the label on an example) has a structure. For a finite set of data structures, the learning algorithm searches for the structured output that minimizes the expected cost $E[c_{h(x)}]$ of example x [44] where cost measures the dissimilarity between the example and a proposed structured output.

An example of a structured-prediction problem is to predict a DNA-binding site on the protein where an example comprises an arbitrary set of residues on the surface and the corresponding label is some graph structure representing the binding site. For instance, this graph structure could encode the distances between residues that participate in binding. A number of published works have hit on this formulation but have used some post-processing technique (rather than a proper structured-prediction technique) to incorporate basic structural information [38, 45,46]. The accuracy of these methods ranges from 70% [38] to 89% [46] depending on the post-processing technique.

3. APPLICATIONS IN PROTEIN STRUCTURE AND FUNCTION MODELING

In this section, we expand our review in protein–DNA interactions to other machine learning applications to function and structure prediction. While a full review of machine learning applications in computational biology has been reviewed by others [47], here we focus on several popular problems. In the following, we review machine learning applications to protein function with interactions to RNA, membrane, peptide and other protein, as well as the effect of single amino acid polymorphisms on function and the prediction of protein localization in the cell. In addition, we review applications of machine learning to structure prediction including secondary structure prediction, homology modeling, fold recognition and *ab initio* folding.

3.1 Protein–RNA

Similar to protein–DNA interactions, protein–RNA interactions also perform vital roles in the cell including protein synthesis, viral replication, cellular defense and developmental regulation [145,146]. One major direction in the analysis of protein-RNA interactions is to identify proteins that bind RNA based on features derived from physio-chemical properties of the sequence. A number of published works have focused casting this problem as a binary classification problem using the support vector machines (SVM) classifier to identify proteins that bind RNA [33–35, 51]. Each of these works derived large data sets from the SwissProt database and applied the support vector machines classifier to discriminate protein sequences that bind RNA from all other sequences. Since sequence analysis techniques can identity homologous proteins as having similar function, most of these works reduced the redundancy of the data sets below a certain threshold: <40% [35], <25% [33] and <20% [34] sequence identity, each achieving an accuracy of 92%, 77% and 77%, respectively (the last result combines RNA/DNA).

Other studies have focused on identifying surface residues that bind RNA. For example, this problem has been cast in the binary classification setting where the data comprises annotated structures gathered in a fashion similar to DNA-binding residue prediction; these works have employed a number of classifiers including neural networks [48,49,52], SVM [53] and Naïve Bayes [54]. This problem has also been cast in the structured-prediction setting, which is decomposed into a binary classification problem (solved by neural networks) followed by post-processing [50]. Likewise, given the imbalance in the number of examples belonging to the positive class versus the negative class, i.e. one positive to five negative, a few works have employed the cost-sensitive binary classification setting to achieve a balanced sensitivity and specificity using SVM [48,49]. These works range in performance depending on the available data. That is, for sequence-based methods [48,50,52,53] the performance ranges from 74% [48] to 86% [50]. Likewise, the accuracy of structure-based methods [49,54] ranges from 85% to 87%, respectively.

3.2 Protein–membrane interactions

Protein–lipid interactions are involved in many crucial cellular processes including signaling, a critical component of every cell. A large number of cytosolic proteins bind reversibly to the membrane (specifically or non-specifically) in order to perform their function; e.g. they bind to membrane to meet a signaling partner. This can reduce a three-dimensional search for a binding partner to a two-dimensional search on the membrane surface; moreover, this search is further restricted to certain lipids according to specific binding interactions. Interest has grown considerably in recent years in one such group of proteins, known as peripheral membrane-binding proteins, which localize to the membrane in order to find their binding partners [147–149]. These membrane-binding proteins reversibly bind lipids in the membrane using a number of mechanisms including specialized domains or just a specialized surface area.

Previous work cast the problem of predicting a protein to bind lipid in the binary classification setting and applied SVM to solve this problem. Specifically, Bhardwaj et al. [55] constructed a data set of protein structures carefully annotated to bind (or not to bind) membranes in a reversible fashion. Each protein was translated into a feature representation using both structure and sequence-based characteristics. The prediction of the membrane binding behavior of four previously uncharacterized C2 domains were validated by experiments. This work achieved over 90% accuracy in discriminating peripheral membrane-binding proteins; it was further improved to 93% accuracy in a later work [3]. In other work, Lin et al. [56] selected a subset of protein sequences from the SwissProt [150] database comprising a generalized class of proteins that bind (and not bind) lipids. They built an SVM model using features derived from physio-chemical properties of the sequences. Their method is evaluated for both full length sequences and sequence domains achieving 86.8% and 89% sensitivity (probability an example is predicted positive given it is positive), respectively.

3.3 Protein–protein interactions

Proteins interact with many polymers (including themselves) in the cell and protein-protein interactions represent a majority of cellular activity. Interactions between proteins are more complex than interactions between protein–DNA, –RNA or –membrane given the wider range of shapes and possible interactions. Nevertheless, a number of successful approaches have characterized protein-protein interactions by motifs [102], sequence conservation [96–100], structural properties [94,151,152] and hot spots [93,153] (a few select residues on the surface that completely characterize binding). For instance, one approach uses a structured-prediction technique (a probabilistic network similar to Bayesian networks) to search for important motifs based on known interactions [102]; this approach was tested over the MIPS [154] and DIP [155] databases. Other approaches have cast this problem in the binary classification setting using support vector machines [96–98,101] and neural networks [94,99,100] over protein–protein complexes culled from the protein data bank (PDB) [156]. While many of these approaches represent the proteins using sequence conservation and structure-based features [96–98,100] with a performance near 77% accuracy (98% for homodimers alone [99]), a few approaches use purely structure-based features [94,101] achieving 44% accuracy. These latter approaches point out that while conservation is a powerful feature, it often leads to unreliable results [101]. Besides conservation and structure, a recent approach derives features from sequence alone to represent proteins and trained a neural network to discriminate interaction hot spots [95] (rather than full interaction patches); this work claims 89% accuracy in discriminating hot spots using sequence alone.

3.4 Protein–peptide interactions

Immune response is an important cellular process mediated by protein–peptide interactions; specifically, the interaction between major histocompatibility complex (MHC) class I molecules and short pathogenic peptides has been extensively studied to better aid vaccine design for pathogens, autoimmune and cancer [157]. The source of data for most published works comes from three sources: the SYF-PEITHI [158] database, the MHCPEP [159] database and laboratory experiments [135,144]. The representation of the peptides is less complex than for proteins; usually the identity and/or physio-chemical properties of a residue at each position is sufficient. For this task, the goal of the learning algorithm is threefold: to correctly classify peptides binding (classification) [133–141], order binding by affinity (ranking) [138–141,143,160] and assign a reasonable affinity (regression) [141–144]. The learning algorithms applied in the classification and ranking settings include boosting [141], support vector machines [133,138–140,143,160], neural networks [134,135], decision trees [137] and hidden Markov models [136] while only support vector machines [142–144] and boosting were applied to the regression setting. One benchmark [141] suggests boosting Gaussian mixture models performs best in terms of area under the ROC, which achieved 0.976 compared to SVMHC [140], which achieved 0.947 on the same data set.

3.5 Subcellular localization

Knowing the location of a protein in a cell helps to narrow its possible functional characteristics and thusly guide experimental strategies [161]. Subcellular localization is a classic multi-class classification problem where a protein sequence is assigned to one of four to sixteen compartments depending on the problem addressed: single organism, single process or all organisms. There seem to be few consistent data sets in subcellular localization literature where many researchers choose to create their own from the SwissProt. Indeed, there has been considerable research in formulating the subcellular localization problem in the multi-class classification setting and this has been extensively reviewed for neural networks, hidden Markov models, self-organizing map and support vector machines [162]. In more recent research, this problem has been tackled by multi-class extensions of support vector machines (SVM) [103–119]. The two most popular multi-class extensions used in subcellular localization are one-versus-one [103,105,108,110–112,

114,119] and one-versus-all [104,106,107,113–118]. That is, in one-versus-one a bi-nary classifier is trained for every pair of classes and the final prediction belongs to the class with the highest number of votes; in one-versus-all, a binary classifier is trained on the full data set where one class is selected to be positive and all other classes are negative; the final prediction is the class predicted positive when all others are negative. In both of these extensions, a set of predictions may result in a tie between two or more classes; thus, many of the proposed methods estimate a vector measuring probability this example belongs to a particular class and use this probability vector to break any ties [104–107]. Indeed, the probability estimation problem is special case of regression where the training examples have probabilities $\{0,1\}$ and predictions are probability estimates that attempt to reach these target values. Nevertheless, there are a number of other methods to extend binary classifiers to multi-class classification tasks; one such method is hierarchical decomposition [109]. This approach to multi-class tasks leverages extra information often available in many problems. For instance, when classifying whether a protein is located in one of three compartments: the cytoplasm, mitochondria or chloroplast, we can arrange these categories in a hierarchy, where classifying cytoplasm/non-cytoplasm is at the root and mitochondria/chloroplast is the branch followed when non-cytoplasm is predicted. This problem has also been cast in the cost-sensitive (a special case of importance-weighted) multi-class classification setting in order to deal with the large discrepancy in the number of examples belonging to each class [114]. The representation of sequence for many approaches to subcellular localization includes amino acid composition [103–112], conservation [104,106–112], sequence order [105,110,113,114,119], physio-chemical properties [105,108,114–118], presence of motifs [106,107,109], secondary structure prediction [106,109] and accessible surface area prediction [106]. Given the rather large number of features, many of which are irrelevant, a number of approaches utilize feature selection [106,115,118] or train classifiers on sub groups of features and combine the subgroup predictions using another classifier [105,107,110,112]. It is hard to estimate the performance in subcellular localization as there is no definitive benchmark; furthermore there are a number of subproblems.

3.6 Single amino acid polymorphisms

With the completion of the human genome project, attention has shifted to human genomic variation. One type of variation that has received significant interest of late is the single nucleotide polymorphism (SNP). With an average density of 1 in 300 base pairs, SNPs account for a good deal of the individuality and diversity in the human population [163–166]. A SNP that causes an amino acid substitution in the protein product is known as a non-synonymous SNP (*ns*SNP) or also as a single amino acid polymorphism (SAP). Indeed, SAPs account for about 50% of the genetic diseases caused by SNPs [167]. To this end, large scale efforts such as the HapMap project [163] have accumulated considerable SAP-related data in databases such as dbSNP [168]. However, these high-throughput experiments fail to fully characterize SAPs in terms of disease association. Furthermore, the underlying mechanisms that lead disease are poorly understood. Thus, computational efforts have been employed to classify a given SAP in terms of its disease association. Machine learning approaches have cast this problem in the binary classification setting and subsequently solved the problem with a variety of algorithms including Bayesian networks [57], neural networks [58], decision tree [59,60], random forests [61] and SVM [60–64]. The best published results in this problem achieve 82% accuracy using a number of powerful features including structural neighbor profile, secondary structure and conservation [62]. However, the data set used in each study is different, thus preventing a straightforward comparison among those works.

3.7 Protein secondary structure prediction

Predicting the secondary structure from sequence has been a long studied problem [169,170]. Most currently successful methods employ evolutionary information in the form of position specific profiles, which rely on fast, accurate iterative search tools such as PSI-BLAST [121,123,124,127–132]. Other techniques rely on physio-chemical representations [126], multiple sequence alignments [120], and meta-learning (combining outputs of other methods) [125,171]. The secondary structure prediction problem can be viewed as both a multi-class and a structured-prediction problem. Firstly, to handle the multi-class problem, many approaches use multi-class learners such as neural or Bayesian networks [123–125,128,129], multi-class [172] support vector machines [130,131], and multi-class extensions of support vector machines such has one-versus-one [126,131] and one-versus-all [120–122,131,132]. Secondly, whether a particular residue forms a helix, sheet or coil to a large extent depends on its neighbors; this means each example residue is dependent on other example residues, a typical structured-prediction task. To this end, most successful methods use two stages where the first stage makes a multi-class prediction and the second stage utilizes this and the neighboring predictions to determine the best class assignment [127–132]. The current performance of secondary structure algorithms can be found on the EVA web site[1] [173].

3.8 Protein tertiary structure prediction

Protein structure prediction is a central problem in molecular biology and accordingly many techniques have been developed to detect the structural class of a primary sequence. The structure of a protein provides a rich set of features, which can be used to determine the function of the protein. However, the corresponding experimental methods such as x-ray crystallography and nuclear magnetic resonance (NMR) spectroscopy are very time consuming (on the order of months to years) and expensive. Moreover, while there are thousands of protein structures in the PDB, there are still millions of sequences with structures that are yet unsolved.

There are two main classification systems to organize proteins based on their structure: CATH [174] and SCOP [175]. These systems are used to label training data for a number of supervised learning problems found in protein structure prediction. This problem is divided into three subproblems depending on the data

---

[1] http://cubic.bioc.columbia.edu/eva/.

available. Firstly, when a sequence shares more than 35% sequence identity to a sequence with a known structure, a sequence analysis technique is sufficient to assign structure. Secondly, when there is no sequence similarity, a similar structure may exist and can be found with a fold assignment algorithm. Finally, the structure can be estimated by *first principles* using a search (or sampling) algorithm in conjunction with some measure of the native state. The performance of the techniques listed below are evaluated every two years in a double blind competition called the critical assessment of techniques for protein structure prediction (CASP[2]).

3.8.1 Homology detection

Many successful techniques have been developed to find similar sequences [176, 177], which generally have the same structure. However, when the sequence similarity drops below a certain threshold (arguably 40% sequence identity), such techniques fail to find sequences that share a similar structure. When there is no result with more than 40% sequence identity, a machine learning technique may then be employed to find a template sequence (with structure). One approach casts this problem in the ranking setting and utilizes the efficient representation of kernel classifiers [66–69,71]. Another approach, called semi-supervised learning, leverages the large amount unlabeled sequence data to build a more accurate model similar to PSI-Blast [72–74]. Yet another approach uses sequence-structure correlations [75,76] or motifs [77] as features in conjunction with kernel methods. Likewise, this problem can be posed in the multi-class classification setting [65] where a sequence is assigned to one of a finite set of families. And finally, this problem has also been cast in the multiple-instance learning setting where an arbitrary number of motifs found in the sequence are represented as instances belonging to the sequence bag. Thus, if a protein sequence belongs to a particular fold (e.g. TrX or Thioredoxin) then one motif must match this fold [78]. This type of problem setting works especially well for folds such as the TrX domain where the existence of a single motif (in conjunction with other features) is enough to classify the protein as having this domain.

3.8.2 Fold assignment

The fold assignment problem matches a sequence to a structural fold irrespective of sequence similarity. One traditional method, called protein threading [178–180], threads a protein sequence along a structure and evaluates some measure of "goodness." In other words, this problem requires the algorithm to search for the best alignment between the query sequence and the target structure. Select published work has focused on selecting the best alignment in the binary ranking setting. Indeed, this work has focused on learning a better scoring function using neural networks over a set of decoys where near native decoys serve as positive examples and other decoys as negative [86–89]. Likewise, the problem of selecting the best alignment has been cast as a regression problem where each training example is labeled with its distance to native [90]. Other work has focused on the alignment as a complex object; in this case, the structured-prediction setting is the

---

[2] http://predictioncenter.gc.ucdavis.edu/.

appropriate machine learning problem. For example, Yu et al. [91,92] attacked this problem using a structural SVM algorithm [181].

The previously described threading algorithm can be viewed as a generative machine learning technique. A generative technique requires only a single example whereas a discriminative technique requires a set of both positive and negative examples. This problem has been cast in the binary classification setting [79] and used the same kernel classifier for remote homology detection [74]. However, this setting does not exclusively assign a sequence to a single fold. Indeed, this problem is more naturally formulated into the multi-class classification setting. Dubchak et al. [81] investigated binary neural network classifiers extended to multi-class classification by one-versus-all [182]. This work was further extended to and compared with one-versus-one [183] using both neural networks and SVM [80] as well as jury voting [82]. A similar approach leverages SVM with a more efficient multi-class representation, DDAG [184], as well as a more biologically relevant feature representation [83,84]. Finally, the SVM classifier was also extended to multiple classes using error-correcting output codes [182], which generally makes a better transform from binary to multi-class [85].

3.8.3 *Ab initio* prediction

When a match between a sequence and structure template cannot be found, an *ab initio* or protein folding [185,186] approach is taken to find the native conformation. This approach can be divided into two steps where the first step samples a conformation and the second step evaluates the energy of the conformation such as using a statistical potential [187]. Finding the best conformation translates to finding the lowest energy conformation. Indeed, this problem has been cast in the binary ranking setting and solved using neural networks [86–89]. While a binary classifier indicates whether the target conformation is near native, an estimate of its distance from native is more useful; to this end, the problem of selecting the best conformation has been cast as a regression problem where each training example is labeled with its distance to native [90]. Likewise, the problem of selecting the native state can also be cast in the structured-prediction setting. One such approach attempts to find the native conformation of residue side-chains using a graph-based model [93].

4. DISCUSSION AND FUTURE OUTLOOK

Machine learning continues to become more and more popular in biomolecular modeling. In this chapter, we described common formalisms to cast modeling applications into machine learning problems. Subsequently, we reviewed a number of recent biomodeling applications found in both literature and our own work, which ranged from structure to localization to interaction to function prediction. Although machine learning has been shown to be quite useful in recent applications, it is still in the early stages and has yet been developed into one of the main tools for computational biophysics. Many of the previous works rely on direct application of available machine learning software in conjunction with simple features. Future progress will go beyond the straightforward application of classification software packages; it will also go beyond the investigation of simple features.

The current focus and development of machine learning for biomolecular modeling can be grouped into four categories: algorithm assessment, algorithm development, knowledge mining and feature representation. That is, in most if not all areas, it is unclear which protocol (algorithm, feature, etc.) performs best given the widely different data sets, validation techniques and evaluation metrics. Likewise, there are many interesting problems beyond simple classification; however, a majority of available algorithms only handle (or perform well in) binary classification. Moreover, the model learned has more value then just a black box predictor; the rules learned by the model can be just as important if not more so. Finally, an overwhelming amount of work has focused on the extracting features from protein sequences and only recently has work begun to tackle protein structure, although it is widely believed that structure will yield better performance.

One of the most crucial components missing in current machine learning work is a unified assessment of performance in specific domains. One of the cornerstones in laboratory experiments is reproducibility and too many machine learning papers are not reproducible. This problem could be rectified in three simple ways. First, published work should be restricted to algorithms in the public domain. While few published works violate this first rule in terms of the machine learning algorithm, the full protocol from feature calculation to parameter selection is often not in the public domain. Since not every minor detail is covered in the publication, this can make it difficult to accurately reproduce an experiment. Second, published work should submit all relevant data sets (features, labels, objects) to a public repository. This incredibly simple step is rarely taken and results in later articles creating new data sets rather than reconstructing old data sets for a fair, unbiased comparison. Third, published work should include every relevant metric and plot to the problem being solved. Indeed, such metrics and plots are cheap for the original researcher to calculate but relatively expensive for other researchers. Moreover, a missing metric may create the impression that there is something to hide. In sum, the current performance of many proposed approaches to biomolecular modeling is largely unknown; at least until one organizes a competition for the particular area such as CASP for protein structure prediction.

Another important area of concentration is the development of new machine learning algorithms and software implementing such algorithms. That is, referring to Table 3.2, a vast majority of applications rely on binary classification (or ranking reduced to binary classification) and (not shown in the table) most of these works rely on support vector machines (SVM). Given the number of available, mature implementations of SVM, it is not surprising that SVM has dominated biomolecular modeling applications. However, SVM is not suited to every problem domain and it is also not the simplest nor the most efficient algorithm available. Thus, in order to both explore more algorithms and more machine learning problem formulations, new algorithms and implementations are necessary to encourage research in new directions toward more sophisticated problems. To this end, the Journal of Machine Learning Research has started a new publication track devoted

Table 3.2   Learning formulations in problem domains

Problem	BIN	MUL	RNK	REG	IWC	MIL	STRUCT
Protein–DNA	[26–31] [33–37,39]	–	[3,32]	[42]	[43]	–	[38,45,46]
Protein–RNA	[33–35] [51–54]	–	–	–	[48,49]	–	[50]
Protein–Mem.	[55,56]	–	–	–	–	–	–
Protein–SAP	[57–64]	–	–	–	–	–	–
Protein–Homol.	–	[65]	[66–77]	–	–	[78]	–
Protein–Fold.	[74,79]	[80–85]	[86–89]	[90]	–	–	[91,92]
Protein–AbIn.	–	–	[86–89]	[90]	–	–	[93]
Protein–protein	[94–101]	–	–	–	–	–	[102]
Subcellular loc.	–	[103–119]	–	–	–	–	-
Secondary struct.	–	[120–126]	–	–	–	–	[127–132]
Peptide-binding	[133–137]	–	[138–140]	[141–144]	–	–	-

BIN: Binary classification. MUL: Multi-class classification. RNK: Ranking. REG: Regression. IWC: Importance-weighted classification. MIL: Multiple-instance Learning. STRUCT: Structured-prediction.

to the development of new machine learning workbenches (collections of machine learning algorithms) [188]. In our own work, we have developed an open source machine learning workbench [189] aimed at making more algorithms accessible to other users.[3] Moreover, this workbench attempts to standardize metrics and validation to facilitate a unified assessment among its users.

Most work has focused on building an accurate model but fails to properly investigate the rules learned from that model. These rules are arguably more important than the predictions they make, in that, they capture the essence of the problem often buried deep within the data. If we knew the important rules necessary to make accurate predictions then machine learning would be of little use. Instead, we rely on machine learning to mine important relationships within the data, which are then used to make an accurate prediction. Few published works have dealt with this difficult problem due to, in a large part, the black box nature of many powerful machine learning algorithms e.g. SVM. Less powerful machine learning algorithms, e.g. decision trees, often build more interpretable models at the expense of prediction performance, which results in less interesting rules being found. In our own work, we have investigated applying the alternating decision tree [190] algorithm to investigate features important in protein–DNA interactions since it both performs well and gives an interpretable model. We validated important relationships found in this model with experimentally derived structures (unpublished work).

Finally, the real value in many papers is derived from the features developed in the search for a meaningful representation of protein sequences and structure. However, too many papers focus on mechanistic feature representations such as

---

[3] http://proteomics.bioengr.uic.edu/malibu/index.html.

sequence composition, transition frequency, and conservation score rather than more biologically meaningful features such as electrostatic or hydrophobic interactions. This is symptomatic of the lack of intuitive software available in the public domain. In general, the search for meaningful features lumbers down two roads: deriving meaningful characteristics of and similarities between examples. The former focuses on extracting important characteristics such as electrostatic patches [26], hydrophobic patches, etc. from a protein. This is particularly important in deriving useful knowledge from the knowledge mining techniques mentioned previously as well as building a better prediction model. The latter focuses on divining a more global measure of similarity between examples, generally in the form of new kernels for algorithms such as SVM; i.e. the string kernel [67], the cluster kernel [73], etc.

Overall, machine learning is a powerful tool in computer science and will find more and more applications in biomolecular modeling. When combined with biochemical and biophysical understanding, this new approach is expected to yield phenomenal advancement in our understanding of protein structures, functions, interactions and localizations in the years to come.

ACKNOWLEDGMENTS

This work is partially supported by NIH P01 AI060915 and the CBC catalyst award (H.L.). R.E.L. acknowledges the support from NIH training grant T32 HL 07692: Cellular Signaling in Cardiovascular System (P.I. John Solaro).

REFERENCES

1. Bhaskar, H., Hoyle, D.C., Singh, S. Machine learning in bioinformatics: A brief survey and recommendations for practitioners. Comput. Biol. Med. 2006, 36, 1104–25.
2. Cios, K.J., Kurgan, L.A., Reformat, M. Machine learning in the life sciences. IEEE Eng. Med. Biol. Mag. 2007, 26, 14–6.
3. Langlois, R., Carson, M., Bhardwaj, N., Lu, H. Learning to translate sequence and structure to function: Identifying DNA binding and membrane binding proteins. Ann. Biomed. Eng. 2007, 35, 1043–52.
4. Bishop, C.M. Neural Networks for Pattern Recognition. Oxford Univ. Press; 1995.
5. Cortes, C., Vapnik, V. Support-vector networks. Mach. Learning 1995, 20, 273–97.
6. Pearl, J. Probabilistic Reasoning in Intelligent Systems: Networks of Plausible Inference. San Francisco, CA, USA: Morgan Kaufmann Publishers Inc.; 1988.
7. Lafferty, J., McCallum, A., Pereira, F. Conditional random fields: Probabilistic models for segmenting and labeling sequence data. In: Proceedings of the 18th International Conference on Machine Learning. San Francisco, CA: Morgan Kaufmann Publishers Inc.; 2001, pp. 282–9.
8. Quinlan, J.R. Induction of decision trees. Mach. Leairning 1986, 1, 81–106.
9. Freund, Y., Schapire, R.E. Experiments with a new boosting algorithm, in: International Conference on Machine Learning, vol. 13, Bari, Italy, 1996, pp. 148–56.
10. Breiman, L. Bagging predictors. Mach. Learning 1996, 24, 123–40.
11. Breiman, L. Random forests. Mach. Learning 2001, 45, 5–32.
12. Shafer, G. Perspectives on the theory and practice of belief functions. Int. J. Approx. Reasoning 1990, 4, 5–6.

13. Tamayo, P., Slonim, D., Mesirov, J., Zhu, Q., Kitareewan, S., Dmitrovsky, E., Lander, E.S., Golub, T.R. Interpreting patterns of gene expression with self-organizing maps: Methods and application to hematopoietic differentiation. Proc. Natl. Acad. Sci. USA 1999, 96, 2907–12.

14. Brown, M.P.S., Grundy, W.N., Lin, D., Cristianini, N., Sugnet, C.W., Furey, T.S., Ares Manuel, J., Haussler, D. Knowledge-based analysis of microarray gene expression data by using support vector machines. Proc. Natl. Acad. Sci. USA 2000, 97, 262–7.

15. Bystroff, C., Baker, D. Prediction of local structure in proteins using a library of sequence-structure motifs. J. Mol. Biol. 1998, 281, 565–77.

16. Bystroff, C., Shao, Y. Fully automated ab initio protein structure prediction using I-SITES, HMM-STR and ROSETTA. Bioinformatics 2002, 18, S54–61.

17. Bystroff, C., Thorsson, V., Baker, D. HMMSTR: A hidden Markov model for local sequence-structure correlations in proteins. J. Mol. Biol. 2000, 301, 173–90.

18. DeLano, W.L. The PyMOL User's Manual. Palo Alto, CA, USA: DeLano Scientific; 2002.

19. Luscombe, N.M., Austin, S.E., Berman, H.M., Thornton, J.M. An overview of the structures of protein–DNA complexes. Genome Biol. 2000, 1, 7558–62.

20. Luscombe, N.M., Thornton, J.M. Protein–DNA interactions: Amino acid conservation and the effects of mutations on binding specificity. J. Mol. Biol. 2002, 320, 991–1009.

21. Cajone, F., Salina, M., Benelli-Zazzera, A. 4-hydroxynonenal induces a DNA-binding protein similar to the heat-shock factor. Biochem. J. 1989, 262, 977–9.

22. Buck, M.J., Lieb, J.D. ChIP-chip: Considerations for the design, analysis, and application of genome-wide chromatin immunoprecipitation experiments. Genomics 2004, 83, 349–60.

23. Ren, B., Robert, F., Wyrick, J.J., Aparicio, O., Jennings, E.G., Simon, I., Zeitlinger, J., Schreiber, J., Hannett, N., Kanin, E., Volkert, T.L., Wilson, C.J., Bell, S.P., Young, R.A. Genome-wide location and function of DNA binding proteins. Science 2000, 290, 2306–9.

24. Chou, C.-C., Lin, T.-W., Chen, C.-Y., Wang, A.H.J. Crystal structure of the hyperthermophilic archaeal DNA-binding protein Sso10b2 at a resolution of 1.85 angstroms. J. Bacteriol. 2003, 185, 4066–73.

25. Ruvkun, G.B., Ausubel, F.M. A general method for site-directed mutagenesis in prokaryotes. Nature 1981, 289, 85–8.

26. Bhardwaj, N., Langlois, R.E., Zhao, G., Lu, H. Kernel-based machine learning protocol for predicting DNA-binding proteins. Nucleic Acids Res. 2005, 33, 6486–93.

27. Shanahan, H.P., Garcia, M.A., Jones, S., Thornton, J.M. Identifying DNA-binding proteins using structural motifs and the electrostatic potential. Nucleic Acids Res. 2004, 32, 4732–41.

28. Pellegrini-Calace, M., Thornton, J.M. Detecting DNA-binding helix-turn-helix structural motifs using sequence and structure information. Nucleic Acids Res. 2005, 33, 2129–40.

29. Stawiski, E.W., Gregoret, L.M., Mandel-Gutfreund, Y. Annotating nucleic acid-binding function based on protein structure. J. Mol. Biol. 2003, 326, 1065–79.

30. Ahmad, S., Sarai, A. Moment-based prediction of DNA-binding proteins. J. Mol. Biol. 2004, 341, 65–71.

31. Bhardwaj, N., Langlois, R.E., Zhao, G., Lu, H. Structure based prediction of binding residues on DNA-binding proteins, in: 27th Annual International Conference on the IEEE EMBS, Shanghai, China, 2005, pp. 2611–14.

32. Szilagyi, A., Skolnick, J. Efficient prediction of nucleic acid binding function from low-resolution protein structures. J. Mol. Biol. 2006, 358, 922–33.

33. Yu, X., Cao, J., Cai, Y., Shi, T., Li, Y. Predicting rRNA-, RNA-, and DNA-binding proteins from primary structure with support vector machines. J. Theor. Biol. 2006, 240, 175–84.

34. Fujishima, K., Komasa, M., Kitamura, S., Suzuki, H., Tomita, M., Kanai, A. Proteome-wide prediction of novel DNA/RNA-binding proteins using amino acid composition and periodicity in the hyperther-mophilic Archaeon Pyrococcus furiosus. DNA Res. 2007, 14, 91–102.

35. dong Cai, Y., Lin, S.L. Support vector machines for predicting rRNA-, RNA-, and DNA-binding proteins from amino acid sequence. Biochim. Biophys. Acta 2003, 1648, 127–33.

36. Ahmad, S., Sarai, A. PSSM-based prediction of DNA binding sites in proteins. BMC Bioinformatics 2005, 6, 33.

37. Kuznetsov, I.B., Gou, Z., Li, R., Hwang, S. Using evolutionary and structural information to predict DNA-binding sites on DNA-binding proteins. Proteins: Struct., Funct., Bioinf. 2006, 64, 19–27.

38. Bhardwaj, N., Lu, H. Residue-level prediction of DNA-binding sites and its application on DNA-binding protein predictions. FEBS Lett. 2007, 581, 1058–66.

39. Yan, C., Terribilini, M., Wu, F., Jernigan, R., Dobbs, D., Honavar, V. Predicting DNA-binding sites of proteins from amino acid sequence. BMC Bioinformatics 2006, 7, 262–72.

40. Balcan, M.-F., Bansal, N., Beygelzimer, A., Coppersmith, D., Langford, J., Sorkin, G. Robust reductions from ranking to classification. In: Computational Learning Theory. Lecture Notes in Computer Science, vol. 4539. Berlin: Springer/Heidelberg; 2007, pp. 604–19.

41. Hansch, C., Maloney, P.P., Toshio, F., Muir, R.M. Correlation of biological activity of phenoxyacetic acids with Hammett substituent constants and partition coefficients. Nature 1962, 194, 178–80.

42. Sharon, E., Segal, E. A feature-based approach to modeling protein–DNA interactions. In: Speed, T.P., Huang, H., editors. Research in Computational Molecular Biology. Lecture Notes in Computer Science, vol. 4453. Berlin: Springer/Heidelberg; 2007, pp. 77–91.

43. Ahmad, S., Gromiha, M.M., Sarai, A. Analysis and prediction of DNA-binding proteins and their binding residues based on composition, sequence and structural information. Bioinformatics 2004, 20, 477–86.

44. Daumé III, H., Langford, J., Marcu, D. Search-based structured prediction. Mach. Learn. J., submitted for publication.

45. Tjong, H., Zhou, H.-X. DISPLAR: An accurate method for predicting DNA-binding sites on protein surfaces. Nucleic Acids Res. 2007, 35, 1465–77.

46. Ofran, Y., Mysore, V., Rost, B. Prediction of DNA-binding residues from sequence. Bioinformatics 2007, 23, i347–53.

47. Noble, W.S.. In: Scholkopf, B., Tsuda, K., Vert, J.-P., editors. Kernel Methods in Computational Biology. MIT Press; 2004, pp. 71–92.

48. Wang, L., Brown, S.J. Prediction of RNA-binding residues in protein sequences using support vector machines. In: 28th Annual International Conference on the IEEE EMBS, New York City, USA, 2006, pp. 5830–33.

49. Wang, Y., Xue, Z., Shen, G., Xu, J. PRINTR: Prediction of RNA binding sites in proteins using SVM and profiles. Amino Acids 2008.

50. Jeong, E., Chung, I.-F., Miyano, S. A neural network method for identification of RNA-interacting residues in protein. Genome Inform. 2004, 15, 105–16.

51. Han, L.Y., Cai, C.Z., Lo, S.L., Chung, M.C.M., Chen, Y.Z. Prediction of RNA-binding proteins from primary sequence by a support vector machine approach. RNA 2004, 10, 355–68.

52. Jeong, E., Miyano, S.. In: Cardelli, L., Emmott, S., Cardelli, L., Emmott, S., editors. Transactions on Computational Systems, Biology IV. Lecture Notes in Computer Science, vol. 3939. Berlin: Springer/Heidelberg; 2006, pp. 123–39.

53. Kumar, M., Gromiha, M.M., Raghava, G.P.S. Prediction of RNA binding sites in a protein using SVM and PSSM profile. Proteins: Struct., Funct., Bioinf. 2008, 71, 189–94.

54. Terribilini, M., Lee, J.-H., Yan, C., Jernigan, R.L., Honavar, V., Dobbs, D. Prediction of RNA binding sites in proteins from amino acid sequence. RNA 2006, 12, 1450–62.

55. Bhardwaj, N., Stahelin, R.V., Langlois, R.E., Cho, W., Lu, H. Structural bioinformatics prediction of membrane-binding proteins. J. Mol. Biol. 2006, 359, 486–95.

56. Lin, H., Han, L., Zhang, H., Zheng, C., Xie, B., Chen, Y. Prediction of the functional class of lipid binding proteins from sequence-derived properties irrespective of sequence similarity. J. Lipid Res. 2006, 47, 824–31.

57. Cai, Z., Tsung, E.F., Marinescu, V.D., Ramoni, M.F., Riva, A., Kohane, I.S. Bayesian approach to discovering pathogenic SNPs in conserved protein domains. Hum. Mutat. 2004, 24, 178–84.

58. Ferrer-Costa, C., Orozco, M., de la Cruz, X. Sequence-based prediction of pathological mutations. Proteins: Struct., Funct., Bioinf. 2004, 57, 811–9.

59. Dobson, R., Munroe, P., Caulfield, M., Saqi, M. Predicting deleterious nsSNPs: An analysis of sequence and structural attributes. BMC Bioinformatics 2006, 7, 217.

60. Krishnan, V.G., Westhead, D.R. A comparative study of machine-learning methods to predict the effects of single nucleotide polymorphisms on protein function. Bioinformatics 2003, 19, 2199–209.

61. Bao, L., Cui, Y. Prediction of the phenotypic effects of non-synonymous single nucleotide polymorphisms using structural and evolutionary information. Bioinformatics 2005, 21, 2185–90.

62. Ye, Z.-Q., Zhao, S.-Q., Gao, G., Liu, X.-Q., Langlois, R.E., Lu, H., Wei, L. Finding new structural and sequence attributes to predict possible disease association of single amino acid polymorphism (SAP). Bioinformatics 2007, 23, 1444–50.
63. Karchin, R., Diekhans, M., Kelly, L., Thomas, D.J., Pieper, U., Eswar, N., Haussler, D., Sali, A. LS-SNP: Large-scale annotation of coding non-synonymous SNPs based on multiple information sources. Bioinformatics 2005, 21, 2814–20.
64. Yue, P., Moult, J. Identification and analysis of deleterious human SNPs. J. Mol. Biol. 2006, 356, 1263–74.
65. Ie, E., Weston, J., Noble, W.S., Leslie, C. Multi-class protein fold recognition using adaptive codes, in: International Conference on Machine Learning, vol. 22, Bonn, Germany, 2005, pp. 329–36.
66. Leslie, C., Eskin, E., Cohen, A., Weston, J., Noble, W.S. Mismatch string kernels for discriminative protein classification. Bioinformatics 2004, 20, 467–76.
67. Leslie, C., Eskin, E., Noble, W.S. The spectrum kernel: A string kernel for SVM protein classification, in: Pacific Symposium on Biocomputing, Lihue, HI, 2002, pp. 564–75.
68. Leslie, C., Eskin, E., Weston, J., Noble, W.S. Mismatch string kernels for SVM protein classification. In: Advances in Neural Information Processing Systems, vol. 15, Vancouver, BC, Canada, 2002, pp. 1441–8.
69. Liao, L., Noble, W.S. Combining pairwise sequence similarity and support vector machines for remote protein homology detection. In: International Conference on Research in Computational Molecular Biology, vol. 6, Washington DC, USA, 2002, pp. 225–32.
70. Jaakkola, T., Diekhans, M., Haussler, D. A discriminative framework for detecting remote protein homologies. J. Comput. Biol. 2000, 7, 95–114.
71. Saigo, H., Vert, J.-P., Ueda, N., Akutsu, T. Protein homology detection using string alignment kernels. Bioinformatics 2004, 20, 1682–9.
72. Kuang, R., Ie, E., Wang, K., Wang, K., Siddiqi, M., Freund, Y., Leslie, C. Profile-based string kernels for remote homology detection and motif extraction. J. Bioinform. Comput. Biol. 2005, 3, 527–50.
73. Weston, J., Leslie, C., Ie, E., Zhou, D., Elisseeff, A., Noble, W.S. Semi-supervised protein classification using cluster kernels. Bioinformatics 2005, 21, 3241–7.
74. Rangwala, H., Karypis, G. Profile-based direct kernels for remote homology detection and fold recognition. Bioinformatics 2005, 21, 4239–47.
75. Hou, Y., Hsu, W., Lee, M.L., Bystroff, C. Efficient remote homology detection using local structure. Bioinformatics 2003, 19, 2294–301.
76. Hou, Y., Hsu, W., Lee, M.L., Bystroff, C. Remote homolog detection using local sequence-structure correlations. Proteins: Struct., Funct., Bioinf. 2004, 57, 518–30.
77. Ben-Hur, A., Brutlag, D. Remote homology detection: A motif based approach, in: International Conference on Intelligent Systems for Molecular Biology, vol. 11, 2003, pp. suppl_l i26–33.
78. Scott, S.D., Ji, H., Wen, P., Fomenko, D.E., Gladyshev, V.N. On modeling protein superfamilies with low primary sequence conservation, Technical report UNL-CSE-2003-4, University of Nebraska, 2003.
79. Cheng, J., Baldi, P. A machine learning information retrieval approach to protein fold recognition. Bioinformatics 2006, 22, 1456–63.
80. Ding, C.H.Q., Dubchak, I. Multi-class protein fold recognition using support vector machines and neural networks. Bioinformatics 2001, 17, 349–58.
81. Dubchak, I., Muchnik, I., Mayor, C., Dralyuk, I., Kim, S.-H. Recognition of a protein fold in the context of the SCOP classification. Proteins: Struct., Funct., Genet. 1999, 35, 401–7.
82. Yu, C.-S., Wang, J.-Y., Yang, J.-M., Lyu, P.-C., Lin, C.-J., Hwang, J.-K. Fine-grained protein fold assignment by support vector machines using generalized n-peptide coding schemes and jury voting from multiple-parameter sets. Proteins: Struct., Funct., Genet. 2003, 50, 531–6.
83. Langlois R.E., Diec, A., Dai, Y., Lu, H. Kernel based approach for protein fold prediction from sequence, in: 26th Annual International Conference on the IEEE EMBS, San Francisco 2004, pp. 2885–8.
84. Langlois, R.E., Diec, A., Perisic, O., Dai, Y., Lu, H. Improved protein fold assignment using support vector machines. Int. J. Bioinform. Res. Appl. 2006, 1, 319–34.
85. Melvin, I., Ie, E., Weston, J., Noble, W.S., Leslie, C. Multi-class protein classification using adaptive codes. J. Mach. Learn. Res. 2007, 8, 1557–81.

86. Tan, C.-W., Jones, D. Using neural networks and evolutionary information in decoy discrimination for protein tertiary structure prediction. BMC Bioinformatics 2008, 9, 94.
87. Jiang, N., Wu, W.X., Mitchell, I. Threading with environment-specific score by artificial neural networks. Soft Comput. 2006, 10, 305–14.
88. Chang, I., Cieplak, M., Dima, R.I., Maritan, A., Banavar, J.R. Protein threading by learning. Proc. Natl. Acad. Sci. USA 2001, 98, 14350–5.
89. Lin, K., May, A.C.W., Taylor, W.R. Threading using neural network (TUNE): The measure of protein sequence-structure compatibility. Bioinformatics 2002, 18, 1350–7.
90. Jiao, F., Xu, J., Yu, L., Schuurmans, D. Protein fold recognition using gradient boosting, in: IEEE Computational Systems Bioinformatics Conference, IEEE, Stanford, CA, 2006, pp. 43–53.
91. Yu, C.-N., Joachims, T., Elber, R. Training protein threading models using structural SVMs, in: ICML Workshop on Learning in Structured Output Spaces, 2006.
92. Yu, C.-N., Joachims, T., Elber, R., Pillardy, J. Support vector training of protein alignment models. In: Research in Computational Molecular Biology. Lecture Notes in Computer Science, vol. 4453. Berlin: Springer/Heidelberg; 2007, pp. 253–67.
93. Yanover, C., Weiss, Y. Approximate inference and protein-folding. In: Becker, K.O.S., Thrun, S., editors. Advances in Neural Information Processing Systems, vol. 15. Vancouver, BC, Canada: MIT Press; 2002, pp. 1457–64.
94. Keil, M., Exner, T.E., Brickmann, J. Pattern recognition strategies for molecular surfaces: III. Binding site prediction with a neural network. J. Comput. Chem. 2004, 25, 779–89.
95. Ofran, Y., Rost, B. Protein–protein interaction hotspots carved into sequences. PLoS Comput. Biol. 2007, 3, e119.
96. Bradford, J.R., Westhead, D.R. Improved prediction of protein–protein binding sites using a support vector machines approach. Bioinformatics 2005, 21, 1487–94.
97. Sen, T.Z., Kloczkowski, A., Jernigan, R.L., Yan, C., Honavar, V., Ho, K.-M., Wang, C.-Z., Ihm, Y., Cao, H., Gu, X., Dobbs, D. Predicting binding sites of hydrolase–inhibitor complexes by combining several methods. BMC Bioinformatics 2004, 5, 205.
98. Wang, B., Wong, H.S., Huang, D.-S. Inferring protein–protein interacting sites using residue conservation and evolutionary information. Protein Pept. Lett. 2006, 13, 999–1005.
99. Valdar, W.S., Thornton, J.M. Conservation helps to identify biologically relevant crystal contacts. J. Mol. Biol. 2001, 313, 399–416.
100. Fariselli, P., Pazos, F., Valencia, A., Casadio, R. Prediction of protein–protein interaction sites in heterocomplexes with neural networks. Eur. J. Biochem. 2002, 269, 1356–61.
101. Kufareva, I., Budagyan, L., Raush, E., Totrov, M., Abagyan, R. PIER: Protein interface recognition for structural proteomics. Proteins: Struct., Funct., Bioinf. 2007, 67, 400–17.
102. Wang, H., Segal, E., Ben-Hur, A., Roller, D., Brutlag, D.L. Identifying protein–protein interaction sites on a genome-wide scale. In: Saul, L.K., Weiss, Y., Bottou, L., editors. Advances in Neural Information Processing Systems, vol. 17. Cambridge, MA: MIT Press; 2005, pp. 1465–72.
103. Hua, S., Sun, Z. Support vector machine approach for protein subcellular localization prediction. Bioinformatics 2001, 17, 721–8.
104. Xie, D., Li, A., Wang, M., Fan, Z., Feng, H. LOCSVMPSI: A web server for subcellular localization of eukaryotic proteins using SVM and profile of PSI-BLAST. Nucleic Acids Res. 2005, 33, W105–10.
105. Yu, C.-S., Chen, Y.-C., Lu, C.-H., Hwang, J.-K. Prediction of protein subcellular localization. Proteins: Struct., Funct., Bioinf. 2006, 64, 643–51.
106. Su, E.C.-Y., Chiu, H.-S., Lo, A., Hwang, J.-K., Sung, T.-Y., Hsu, W.-L. Protein subcellular localization prediction based on compartment-specific features and structure conservation. BMC Bioinformatics 2007, 8, 330.
107. Höglund, A., Dönnes, P., Blum, T., Adolph, H.-W., Kohlbacher, O. MultiLoc: Prediction of protein subcellular localization using n-terminal targeting sequences, sequence motifs and amino acid composition. Bioinformatics 2006, 22, 1158–65.
108. Bhasin, M., Raghava, G.P.S. ESLpred: SVM-based method for subcellular localization of eukaryotic proteins using dipeptide composition and PSI-BLAST. Nucleic Acids Res. 2004, 32, W414–9.
109. Nair, R., Rost, B. Mimicking cellular sorting improves prediction of subcellular localization. J. Mol. Biol. 2005, 348, 85–100.
110. Guo, J., Lin, Y., Liu, X. GNBSL: A new integrative system to predict the subcellular location for gram-negative bacteria proteins. Proteomics 2006, 6, 5099–105.

111. Rashid, M., Sana, S., Raghava, G.P. Support vector machine-based method for predicting sub-cellular localization of mycobacterial proteins using evolutionary information and motifs. BMC Bioinformatics 2007, 8, 337.

112. Garg, A., Bhasin, M., Raghava, G.P.S. Support vector machine-based method for subcellular localization of human proteins using amino acid compositions, their order, and similarity search. J. Biol. Chem. 2005, 280, 14427–32.

113. Lei, Z., Dai, Y. An SVM-based system for predicting protein subnuclear localizations. BMC Bioinformatics 2005, 6, 291.

114. Shi, J.-Y., Zhang, S.-W., Pan, Q., Cheng, Y.-M., Xie, J. Prediction of protein subcellular localization by support vector machines using multi-scale energy and pseudo amino acid composition. Amino Acids 2007, 33, 69–74.

115. Huang, W.-L., Tung, C.-W., Huang, H.-L., Hwang, S.-F., Ho, S.-Y. ProLoc: Prediction of protein subnuclear localization using SVM with automatic selection from physicochemical composition features. Biosystems 2007, 90, 573–81.

116. Zhang, Z.-H., Wang, Z.-H., Zhang, Z.-R., Wang, Y.-X. A novel method for apoptosis protein sub-cellular localization prediction combining encoding based on grouped weight and support vector machine. FEBS Lett. 2006, 580, 6169–74.

117. Chen, Y.-L., Li, Q.-Z. Prediction of apoptosis protein subcellular location using improved hybrid approach and pseudo-amino acid composition. J. Theor. Biol. 2007, 248, 377–81.

118. Huang, W.-L., Tung, C.-W., Ho, S.-W., Hwang, S.-F., Ho, S.-Y. ProLoc-GO: Utilizing informative gene ontology terms for sequence-based prediction of protein subcellular localization. BMC Bioinformatics 2008, 9, 80.

119. Ogul, H., Mumcuogu, E.U. Subcellular localization prediction with new protein encoding schemes. IEEE/ACM Trans. Comput. Biol. Bioinform. 2007, 4, 227–32.

120. Hua, S., Sun, Z. A novel method of protein secondary structure prediction with high segment overlap measure: Support vector machine approach. J. Mol. Biol. 2001, 308, 397–407.

121. Kuang, R., Leslie, C.S., Yang, A.-S. Protein backbone angle prediction with machine learning approaches. Bioinformatics 2004, 20, 1612–21.

122. Wang, L.-H., Liu, J., Li, Y.-F., Zhou, H.-B. Predicting protein secondary structure by a support vector machine based on a new coding scheme. Genome Inform. 2004, 15, 181–90.

123. Jones, D.T. Protein secondary structure prediction based on position-specific scoring matrices. J. Mol. Biol. 1999, 292, 195–202.

124. Yao, X.-Q., Zhu, H., She, Z.-S. A dynamic Bayesian network approach to protein secondary structure prediction. BMC Bioinformatics 2008, 9, 49.

125. Montgomerie, S., Sundararaj, S., Gallin, W.J., Wishart, D.S. Improving the accuracy of protein secondary structure prediction using structural alignment. BMC Bioinformatics 2006, 7, 301.

126. Hu, H.-J., Pan, Y., Harrison, R., Tai, P.C. Improved protein secondary structure prediction using support vector machine with a new encoding scheme and an advanced tertiary classifier. IEEE Trans. Nanobioscience 2004, 3, 265–71.

127. Zimmermann, O., Hansmann, U.H.E. Support vector machines for prediction of dihedral angle regions. Bioinformatics 2006, 22, 3009–15.

128. Boden, M., Yuan, Z., Bailey, T.L. Prediction of protein continuum secondary structure with probabilistic models based on NMR solved structures. BMC Bioinformatics 2006, 7, 68.

129. Adamczak, R., Porollo, A., Meller, J. Combining prediction of secondary structure and solvent accessibility in proteins. Proteins: Struct., Funct., Bioinf. 2005, 59, 467–75.

130. Nguyen, M.N., Rajapakse, J.C. Two-stage multi-class support vector machines to protein secondary structure prediction. In: Pacific Symposium on Biocomputing, 2005, pp. 346–57.

131. Nguyen, M.N., Rajapakse, J.C. Multi-class support vector machines for protein secondary structure prediction. Genome Inform. 2003, 14, 218–27.

132. Guo, J., Chen, H., Sun, Z., Lin, Y. A novel method for protein secondary structure prediction using dual-layer SVM and profiles. Proteins: Struct., Funct., Bioinf. 2004, 54, 738–43.

133. Cui, J., Han, L.Y., Lin, H.H., Zhang, H.L., Tang, Z.Q., Zheng, C.J., Cao, Z.W., Chen, Y.Z. Prediction of MHC-binding peptides of flexible lengths from sequence-derived structural and physicochemical properties. Mol. Immunol. 2007, 44, 866–77.

134. Honeyman, M.C., Brusic, V., Stone, N.L., Harrison, L.C. Neural network-based prediction of candidate T-cell epitopes. Nat. Biotechnol. 1998, 16, 966–9.

135. Gulukota, K., Sidney, J., Sette, A., DeLisi, C. Two complementary methods for predicting peptides binding major histocompatibility complex molecules. J. Mol. Biol. 1997, 267, 1258–67.
136. Mamitsuka, H. Predicting peptides that bind to MHC molecules using supervised learning of hidden Markov models. Proteins: Struct., Funct., Genet. 1998, 33, 460–74.
137. Savoie, C.J., Kamikawaji, N., Sasazuki, T., Kuhara, S. Use of BONSAI decision trees for the identification of potential MHC class I peptide epitope motifs. In: Pacific Symposium on Biocomputing, 1999, pp. 182–9.
138. Zhao, Y., Pinilla, C., Valmori, D., Martin, R., Simon, R. Application of support vector machines for T-cell epitopes prediction. Bioinformatics 2003, 19, 1978–84.
139. Riedesel, H., Kolbeck, B., Schmetzer, O., Knapp, E.-W. Peptide binding at class I major histocompatibility complex scored with linear functions and support vector machines. Genome Inform. 2004, 15, 198–212.
140. Dönnes, P., Elofsson, A. Prediction of MHC class I binding peptides, using SVMHC. BMC Bioinformatics 2002, 3, 25.
141. Hertz, T., Yanover, C. PepDist: A new framework for protein–peptide binding prediction based on learning peptide distance functions. BMC Bioinformatics 2006, 7 (Suppl 1), S3.
142. Zhao, C., Zhang, H., Luan, F., Zhang, R., Liu, M., Hu, Z., Fan, B. QSAR method for prediction of protein–peptide binding affinity: Application to MHC class I molecule HLA-A*0201. J. Mol. Graph. Model. 2007, 26, 246–54.
143. Liu, W., Meng, X., Xu, Q., Flower, D.R., Li, T. Quantitative prediction of mouse class i MHC peptide binding affinity using support vector machine regression (SVR) models. BMC Bioinformatics 2006, 7, 182.
144. Bhasin, M., Raghava, G.P.S. Analysis and prediction of affinity of TAP binding peptides using cascade SVM. Protein Sci. 2004, 13, 596–607.
145. Hall, K.B. RNA–protein interactions. Curr. Opin. Struct. Biol. 2002, 12, 283–8.
146. Tian, B., Bevilacqua, P.C., Diegelman-Parente, A., Mathews, M.B. The double-stranded-RNA-binding motif: Interference and much more. Nat. Rev. Mol. Cell Biol. 2004, 5, 1013–23.
147. Cho, W. Membrane targeting by c1 and c2 domains. J. Biol. Chem. 2001, 276, 32407–10.
148. Hurley, J.H., Meyer, T. Subcellular targeting by membrane lipids. Curr. Opin. Cell Biol. 2001, 13, 146–52.
149. Teruel, M.N., Meyer, T. Translocation and reversible localization of signaling proteins: A dynamic future for signal transduction. Cell 2000, 103, 181–4.
150. Consortium, T.U. The universal protein resource (UniProt). Nucleic Acids Res. 2007, 35, D193–7.
151. Lu, H., Lu, L., Skolnick, J. Development of unified statistical potentials describing protein–protein interactions. Biophys. J. 2003, 84, 1895–901.
152. Bordner, A.J., Abagyan, R. Statistical analysis and prediction of protein–protein interfaces. Proteins: Struct., Funct., Bioinf. 2005, 60, 353–66.
153. Bogan, A.A., Thorn, K.S. Anatomy of hot spots in protein interfaces. J. Mol. Biol. 1998, 280, 1–9.
154. Mewes, H.W., Frishman, D., Güldener, U., Mannhaupt, G., Mayer, K., Mokrejs, M., Morgenstern, B., Münsterkötter, M., Rudd, S., Weil, B. Mips: A database for genomes and protein sequences. Nucleic Acids Res. 2002, 30, 31–4.
155. Xenarios, I., Salwínski, L., Duan, X.J., Higney, P., Kim, S.-M., Eisenberg, D. Dip, the database of interacting proteins: A research tool for studying cellular networks of protein interactions. Nucleic Acids Res. 2002, 30, 303–5.
156. Westbrook, J., Feng, Z., Jain, S., Bhat, T.N., Thanki, N., Ravichandran, V., Gilliland, G.L., Bluhm, W., Weissig, H., Greer, D.S., Bourne, P.E., Berman, H.M. The protein data bank: Unifying the archive. Nucleic Acids Res. 2002, 30, 245–8.
157. Shoshan, S.H., Admon, A. Mhc-bound antigens and proteomics for novel target discovery. Pharmacogenomics 2004, 5, 845–59.
158. Rammensee, H., Bachmann, J., Emmerich, N.P., Bachor, O.A., Stevanovic, S. Syfpeithi: Database for mhc ligands and peptide motifs. Immunogenetics 1999, 50, 213–9.
159. Brusic, V., Rudy, G., Harrison, L.C. Mhcpep, a database of mhc-binding peptides: Update 1997. Nucleic Acids Res. 1998, 26, 368–71.
160. Zhang, G.L., Bozic, I., Kwoh, C.K., August, J.T., Brusic, V. Prediction of supertype-specific HLA class i binding peptides using support vector machines. J. Immunol. Methods 2007, 320, 143–54.

161. Eisenhaber, F., Bork, P. Wanted: Subcellular localization of proteins based on sequence. Trends Cell. Biol. 1998, 8, 169–70.
162. Schneider, G., Fechner, U. Advances in the prediction of protein targeting signals. Proteomics 2004, 4, 1571–80.
163. Consortium, I.H. The international HapMap project. Nature 2003, 426, 789–96.
164. Kruglyak, L., Nickerson, D.A. Variation is the spice of life. Nat. Genet. 2001, 27, 234–6.
165. Reich, D.E., Gabriel, S.B., Altshuler, D. Quality and completeness of SNP databases. Nat. Genet. 2003, 33, 457–8.
166. Pastinen, T., Ge, B., Hudson, T.J. Influence of human genome polymorphism on gene expression. Hum. Mol. Genet. 2006, 15, R9–16.
167. Krawczak, M., Ball, E.V., Fenton, I., Stenson, P.D., Abeysinghe, S., Thomas, N., Cooper, D.N. Human gene mutation database—A biomedical information and research resource. Hum. Mutat. 2000, 15, 45–51.
168. Sherry, S.T., Ward, M., Kholodov, M., Baker, J., Phan, L., Smigielski, E.M., Sirotkin, K. dbSNP: The NCBI database of genetic variation. Nucleic Acids Res. 2001, 29, 308–11.
169. Rost, B. Review: Protein secondary structure prediction continues to rise. J. Struct. Biol. 2001, 134, 204–18.
170. Barton, G.J. Protein secondary structure prediction. Curr. Opin. Struct. Biol. 1995, 5, 372–6.
171. Selbig, J., Mevissen, T., Lengauer, T. Decision tree-based formation of consensus protein secondary structure prediction. Bioinformatics 1999, 15, 1039–46.
172. Crammer, K., Singer, Y. On the algorithmic implementation of multiclass kernel-based vector machines. J. Mach. Learn. Res. 2001, 2, 265–92.
173. Rost, B., Eyrich, V.A. Eva: Large-scale analysis of secondary structure prediction. Proteins 2001, 45, 192–9.
174. Orengo, C.A., Michie, A.D., Jones, S., Jones, D.T., Swindells, M.B., Thornton, J.M. CATH—A hierarchic classification of protein domain structures. Structure 1997, 5, 1093–108.
175. Murzin, A.G., Brenner, S.E., Hubbard, T., Chothia, C. SCOP: A Structural Classification Of Proteins database for the investigation of sequences and structures. J. Mol. Biol. 1995, 247, 536–40.
176. Karplus, K., Barrett, C., Hughey, R. Hidden Markov models for detecting remote protein homologies. Bioinformatics 1998, 14, 846–56.
177. Sanchez, R., Sali, A. Evaluation of comparative protein structure modeling by MODELLER-3. Proteins: Struct., Funct., Genet. 1997, 29, 50–8.
178. Skolnick, J., Kihara, D. Defrosting the frozen approximation: PROSPECTOR—A new approach to threading. Proteins: Struct., Funct., Genet. 2001, 42, 319–31.
179. McGuffin, L.J., Jones, D.T. Improvement of the GenTHREADER method for genomic fold recognition. Bioinformatics 2003, 19, 874–81.
180. Xu, Y., Xu, D. Protein threading using PROSPECT: Design and evaluation. Proteins: Struct., Funct., Genet. 2000, 40, 343–54.
181. Tsochantaridis, I., Hofmann, T., Joachims, T., Altun, Y. Support vector machine learning for interdependent and structured output spaces, in: International Conference on Machine Learning 2004, pp. 823–30.
182. Dietterich, T.G., Bakiri, G. Solving multiclass learning problems via error-correcting output codes. J. Artif. Intell. Res. 1995, 2, 263–86.
183. Allwein, E.L., Schapire, R.E., Singer, Y. Reducing multiclass to binary: A unifying approach for margin classifiers. J. Mach. Learn. Res. 2001, 1, 113–41.
184. Platt, J.C., Cristianini, N., Shawe-Taylor, J. Large margin DAGs for multiclass classification. In: Advances in Neural Information Processing Systems, vol. 12, Denver, CO, 1999, pp. 547–53.
185. Bonneau, R., Tsai, J., Ruczinski, I., Chivian, D., Rohl, C., Strauss, C.E.M., Baker, D. Rosetta in CASP4: Progress in ab initio protein structure prediction. Proteins: Struct., Funct., Genet. 2001, 45, 119–26.
186. Kihara, D., Lu, H., Kolinski, A., Skolnick, J. TOUCHSTONE: An ab initio protein structure prediction method that uses threading-based tertiary restraints. Proc. Natl. Acad. Sci. USA 2001, 98, 10125–30.
187. Lu, H., Skolnick, J. A distance-dependent atomic knowledge-based potential for improved protein structure selection. Proteins 2001, 44, 223–32.

188. Sonnenburg, S., Braun, M.L., Ong, C.S., Bengio, S., Bottou, L., Holmes, G., LeCun, Y., Muller, K.-R., Pereira, F., Rasmussen, C.E., Ratsch, G., Scholkopf, B., Smola, A., Vincent, P., Weston, J., Williamson, R. The need for open source software in machine learning. J. Mach. Learn. Res. 2007, 8, 2443–66.

189. Langlois, R.E., Lu, H., Intelligible machine learning with malibu for bioinformatics and medical informatics, in: 30th Annual International Conference on the IEEE EMBS, Vancouver, 2008.

190. Freund, Y., Mason, L. The alternating decision tree learning algorithm, in: International Conference on Machine Learning, vol. 16, Bled, Slovenia, 1999, pp. 124–33.

CHAPTER 4

Modeling Protein–Protein and Protein–Nucleic Acid Interactions: Structure, Thermodynamics, and Kinetics

Huan-Xiang Zhou[*,1], **Sanbo Qin**[*], and **Harianto Tjong**[*]

1. INTRODUCTION

It is now increasingly recognized that proteins function in the context of multi-component complexes. This review aims to cover recent progress in modeling fundamental properties of proteins in their interactions among themselves and

[*] Department of Physics and Institute of Molecular Biophysics, Florida State University, Tallahassee, Florida 32306, USA
hzhou4@fsu.edu (H.-X. Zhou)
[1] Corresponding author.

Annual Reports in Computational Chemistry, Vol. 4
ISSN 1574-1400, DOI: 10.1016/S1574-1400(08)00004-2

with nucleic acids. We pay special attention to computational papers which appeared in the past three years, but experimental papers and earlier computational papers which we find particularly relevant are also discussed.

We focus on four interrelated aspects of protein–protein and protein–nucleic acid interactions. Section 2 deals with building structural models for protein complexes. In Section 3 we present an overview of the various methods for computing contributions to the stability of protein complexes. In Section 4 the focus shifts to the rates of forming protein complexes. Finally in Section 5 we discuss the impacts of protein dynamics on the structures, thermodynamics, and kinetics of protein complexes.

2. BUILDING STRUCTURAL MODELS

As a result of favorable interactions, a protein and its partner(s) will form a stereospecific complex. Under favorable conditions, the structure of this complex can be determined by X-ray crystallography, NMR, or electron microscopy. The structure holds the key to understanding the interactions involved and is the basis for making computations on the stability and rate of complex formation.

In many cases, for practical or technical reasons (as opposed to any fundamental physical reasons), the structures of protein complexes cannot be determined experimentally. If the structure of a protein complex with adequate sequence similarity is available, one can build the structure of a query complex by homology modeling [1–3]. The applicability of homology modeling to protein complexes is still limited because the current structural database provides only a sparse coverage of the protein interaction space.

The general approach which aims to build the structure of a complex, starting from the structures of the unbound partners, is now referred to as docking. A forum that provides a fair and critical assessment of various docking methods is the CAPRI "experiment" (http://www.ebi.ac.uk/msd-srv/capri/) run by Joël Janin. We strongly urge method developers to participate in CAPRI, and at the minimum, use the CAPRI targets as a test set. Interested readers can find the latest progress report on CAPRI in a special issue of Proteins (Vol. 69, Issue 4, December 2007).

In general, docking methods aim to maximize the shape and/or physiochemical complementarity between binding partners through generation of large sets of possible poses. Both the sampling of relevant poses and the discrimination of near native poses from the large number of non-native alternatives present significant challenges. The task becomes even more daunting when complex formation is accompanied by rearrangement of loops or relative movement of domains. In our (admittedly biased) opinion, a fruitful approach is to make use of any experimental information available on the interaction [4–7]. Interaction sites can also be predicted by various bioinformatics approaches (for a recent review, see [8]), and from a set of known interfaces by screening [9,10].

We briefly mention two related subjects. For obvious reasons, the interfaces of protein complexes have been a target for developing drug molecules. This subject has been reviewed in this series [11]. In addition, it has now become possible to design de novo complexes, either by modifying a monomeric protein into a dimeric form [12] or by grafting from an unrelated protein complex [13].

3. PREDICTION OF BINDING AFFINITIES

The stability of protein complexes is measured by the binding constant (K_a). Experimentally determined values of K_a span over 10 orders of magnitude (see Figure 4.1). It is clear that no simple correlations exist between structures of protein complexes and their binding affinities. General approaches to calculating binding affinities have been reviewed [14]. Here we focus on aspects specific to protein–protein and protein–nucleic acid complexes.

The binding constant is given by [14,15]:

$$K_a = \int_\Gamma dr\, d\omega\, e^{-W(\mathbf{r},\omega)/k_B T} \tag{1}$$

where $W(\mathbf{r}, \omega)$ is the potential of mean force of for the interaction between a protein and its partner at a relative separation $\mathbf{r}$ and relative orientation ω, $k_B T$ is thermal energy, and Γ denotes the region of configurational space defining the bound state. Contributing factors to $W(\mathbf{r}, \omega)$ include hydrophobic and electrostatic interactions, and the change in conformational entropy of the binding partners upon complex formation. Typically, computations aim to predict the change in the binding free energy, $-k_B T \ln K_a$, e.g., due to a point mutation.

3.1 Electrostatic contribution

It is well understood that hydrophobic interactions make favorable contributions to binding. However, the effects of electrostatic interactions are subtle. Neglecting conformational changes, the electrostatic contribution is given by

$$W_{el} = G_{el}(AB) - G_{el}(A) - G_{el}(B) \tag{2}$$

where G_{el} is the electrostatic free energy of each subunit (A or B) or the complex (AB), which can be calculated by solving the Poisson–Boltzmann (PB) equation. The subtlety of the electrostatic contribution can be appreciated by decomposing it into two components: the desolvation cost W_{desol} and the solvent-screened interaction energy W_{int} (Figure 4.2). To obtain W_{desol}, the electrostatic solvation energy of each subunit is calculated twice, first by itself and then in the presence of its partner, which has its partial charges zeroed out. The difference in the results between these two calculations gives the desolvation cost for that subunit, and adding the corresponding quantity for its partner gives W_{desol}. The difference between W_{el} and W_{desol} comes from the interactions between the partial charges of the two subunits in the solvent environment.

It is clear that W_{desol} opposes binding. W_{int} will favor binding when the charges on the two subunits have complementary charge distributions, which should be

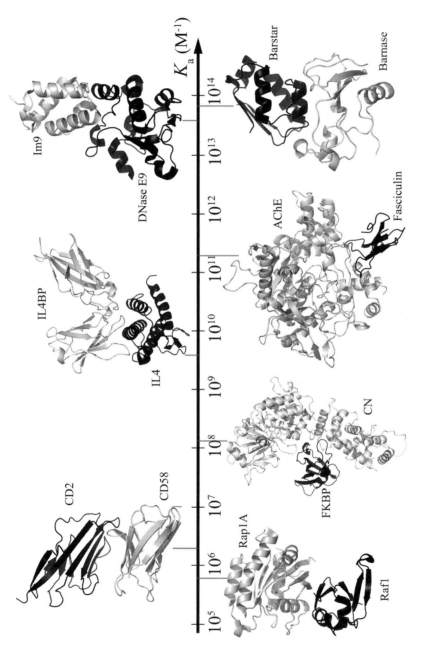

FIGURE 4.1 The spectrum of protein affinities. The locations of seven protein–protein complexes within the spectrum, along with their structures, are shown. Adapted from Dong and Zhou [17].

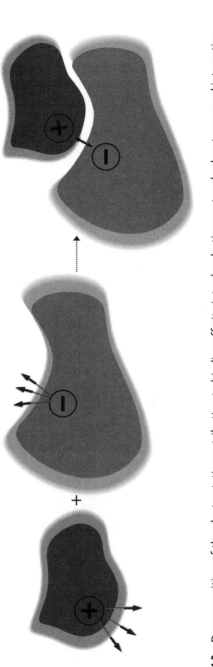

FIGURE 4.2 Decomposition of the electrostatic contribution to binding affinity into desolvation cost and solvent-screened interaction. Interactions of protein charges with the solvent (represented by shadows around binding molecules) are indicated by outgoing arrows. Upon binding, the binding molecules are desolvated within their interface and charge–charge interactions, as indicated by a double-headed arrow, emerge.

true in general. There W_{el} consists of two opposite components. Whether electrostatic interactions make a net positive or net negative contribution to binding rests on the balance between the two components. In particular, the balance is very sensitive to how the boundary between the protein low dielectric and the solvent high dielectric is precisely specified. As shown on a large number of protein–protein and protein–RNA complexes [16–19], when the dielectric boundary is chosen as the molecular surface (MS), as is often done in the literature, W_{desol} outweighs W_{int}, leading to net destabilization. However, when the dielectric boundary is switched to the van der Waals (vdW) surface, the situation is reversed and electrostatic stabilization is now predicted.

How can one then decide on the choice of the dielectric boundary? One possibility is to benchmark PB calculations against explicit-solvent molecular dynamics (MD) simulations. Most of such efforts have been limited to small solute molecules [20–22]. However, it has been shown that the difference between MS and vdW results for electrostatic solvation energies depends on solute size [23]. Therefore parameterization on small solutes (either against explicit-solvent MD results or against experimental data) may not be reliable for calculating electrostatic contributions to protein–protein and protein–nucleic acid binding.

One can benchmark PB calculations directly against experimental data on protein–protein and protein–nucleic acid binding affinities. Potentially one type of useful data is the dependence of binding affinities on salt concentration. The screening of electrostatic interactions by salts can be captured by the PB equation (it should be mentioned that salts can also specifically bind to proteins and nucleic acids; such specific salt effects require special treatment). Unfortunately, it has been found that the screening effects predicted by MS and vdW calculations are essentially identical and thus cannot discriminate between the two choices of the dielectric boundary [16,18]. On the other hand, effects of mutations involving charged or polar residues have been found to have discriminating power, with experimental data favoring the vdW surface as the choice for the dielectric boundary [16–18]. Experimental data for mutational effects on binding affinity continue to accumulate in the literature [24,25], providing opportunities for comprehensive benchmarking of PB calculation protocols.

In the literature, the MS is still widely chosen as the dielectric boundary. The difference between this choice and the vdW surface is that, according to the latter protocol, the many crevices in the protein interior are treated as part of the solvent high dielectric. These crevices are not accessible to a spherical solvent used in defining the MS, and hence their being treated as part of the solvent dielectric is perceived as unrealistic or undesirable. However, this perception is open to question. Water molecules can access protein interiors, as demonstrated by many protein X-ray structures with water occupying interior positions, by the observation of positionally disordered water molecules in a hydrophobic cavity of interleukin 1β [26] (Figure 4.3), and by molecular dynamics simulations [27]. In proteins like myoglobin and acetylcholinesterase (featuring a deeply buried active site connected to the exterior only through a narrow gorge), access by small molecules like water, made possible by the dynamics of the proteins, is essential for biological functions. We suggest that the vdW protocol provides a way to ac-

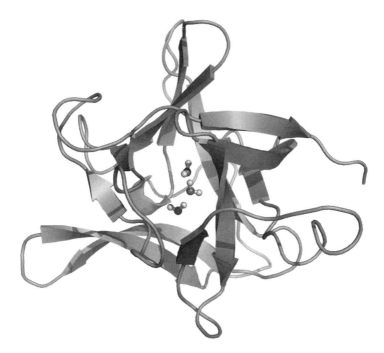

FIGURE 4.3   The presence of water molecules inside a hydrophobic cavity of interleukin 1β. The cavity is separated from the bulk solution according to the MS criterion but connected to the bulk solution according to the vdW criterion. When the three water molecules are moved from separate positions in the bulk solution to the configurations shown inside the cavity, the MS protocol predicts an increase of 0.9 kcal/mol in electrostatic free energy whereas the vdW protocol predicts a decrease of −2.2 kcal/mol.

count for water access to protein interiors. Failure to account for this important property is perhaps the cause for overprediction of pK_a shifts by the MS protocol (which is often "corrected" by increasing the protein dielectric constant to 20). In principle the vdW protocol can be mimicked by the MS protocol with appropriately reduced atomic radii. However, it has been found the precise amount of radius reduction varies from protein to protein and thus mimicking one protocol by the other appears to be a futile exercise [23]. We will come back to the debate between MS and vdW in Section 4.3.

The generalized Born (GB) model has been developed as a fast substitute of the PB equation [28–31]. The GB model can be tailored to match PB results for electrostatic solvation energies obtained by either the MS or the vdW protocol. The errors of GB results in reproducing the PB counterparts are at least of the order of typical mutational effects on binding affinities. Therefore caution should be exercised when applying the GB model to calculate mutational effects.

There is also progress in the opposite direction, i.e., toward more accurate modeling of electrostatic effects, by accounting for electronic polarization via quantum mechanical treatments [32,33]. Such treatments have not been used to directly predict the effects of mutations on the binding free energy, but it is already clear that electronic polarization can significantly influence electrostatic contributions to binding.

Comparing PB or GB calculations against experimental data for mutational effects on binding affinity is premised on the assumption that the mutational effects are assumed to be dominated by electrostatic contributions. That is, possible contributions by hydrophobic interactions and by changes of conformational entropy are not taken into consideration.

3.2 Other contributions

The limitations listed in the last paragraph are dealt with by the molecular mechanics Poisson–Boltzmann surface area (MM-PBSA) method [34,35]. Like before, the electrostatic contribution is calculated by solving the PB equation, but now the hydrophobic contribution is also calculated (as a linear function of the buried surface area), as is the change in conformational entropy [from (quasi)harmonic analyses of conformational fluctuations]. (There is also a version in which PB is replaced by GB [36].) In recent applications, this method has been used to validate homology models of protein–protein complexes [37] and to elucidate molecular bases of promiscuity and selectivity of protein–protein binding [38,39]. Extensions include using different protein dielectric constants for different types of mutated residues [40] and a simplified way of calculating the change in conformational entropy [41].

Another approach, called linear interaction energy (LIE) [42], is somewhat similar to the MM-PBSA method. Here the electrostatic and van der Waals interactions energies of the residue under mutation with its surroundings are calculated in MD simulations of the complex and of the subunit. The changes of these two energies upon binding are then used in a linear regression against a training data set. In this context, we note that many other quantities, including the various components of MM-PBSA calculations [43,44] and physical descriptors such the number of interfacial salt bridges and hydrogen bonds [45] have been used for linear regression. A limitation of all these methods is the requirement of a training data set.

Particularly worth mentioning are computational redesigns which have led to increased protein–protein binding affinities [46–48] or specificity [49]. These redesign methods use physically-based energy functions. These functions involve a large number of parameters, but these parameters are pre-fixed and not adjusted for predicting binding affinities.

4. PREDICTION OF BINDING RATES

The critical role of protein–protein and protein–nucleic acid binding rates is obvious in biological processes in which speed is of the essence [50]. One such example is provided by the purple cone snail, which captures its prey with remarkable efficiency and speed through releasing a polypeptide toxin that rapidly binds to a potassium channel [51]. Compelling arguments can be made for the biological roles of rapid binding in general [52]. In particular, when several proteins compete for the same receptor or when one protein is faced with alternative pathways, kinetic control, not thermodynamic control, dominates for much of the time. Differences in binding rate between related proteins may serve as an additional mechanism for specificity. In short, rapid binding may be as important as high affinity in the proper functioning of proteins. In designing drugs targeting protein–protein interactions, both binding affinity and binding rate may have to be taken into consideration.

4.1 Overview of protein binding rates

Experimentally observed binding rates cover a wide spectrum, from $< 10^3$ $M^{-1} s^{-1}$ to $\sim 10^{10}$ $M^{-1} s^{-1}$ (Figure 4.4). To gain an overview on the wide variation in binding rates, we have considered the binding of two proteins (A and B) as going through an intermediate state (A*B), in which the two proteins have near-native separations and orientations [53,54]. We refer to the intermediate state as the transient complex (its precise specification is given below; a related but more loosely defined term is encounter complex). It is of interest to note that NMR has enabled visualization of the transient complex [55]. From the transient complex, conformational rearrangement can lead to the native complex (AB). Accordingly we have the kinetic scheme

$$A + B \xrightleftharpoons[k_{-D}]{k_D} A * B \xrightarrow{k_c} C .$$ (3)

The overall binding is

$$k_a = \frac{k_D k_c}{k_{-D} + k_c}$$ (4)

which is bounded by the diffusion-controlled rate, k_D, for reaching the transient complex. This limit is reached when conformational rearrangement is fast (i.e., $k_c \gg k_{-D}$), leading to

$$k_a \approx k_D.$$ (5)

At the other end of the spectrum, conformational rearrangement is rate-limiting (i.e., $k_c \ll k_{-D}$), and

$$k_a \approx k_c k_D / k_{-D} \equiv k_R.$$ (6)

Note that k_D / k_{-D} is the equilibrium constant for forming the transient complex.

In the transient complex the two protein molecules must satisfy translational/rotational constraints, which severely hinder the diffusion-controlled rate k_D. In the absence of any biasing force, theoretical estimates put the basal value, k_{D0}, in the range of 10^5 to 10^6 $M^{-1} s^{-1}$ [56–58]. In particular, antibody-protein binding rates are typically observed in this narrow range [59–61]. The value 10^5 $M^{-1} s^{-1}$ thus marks the start of the diffusion-controlled regime (Figure 4.4). A rate much lower than 10^5 $M^{-1} s^{-1}$ is an indication that conformational change plays a significant role in the association.

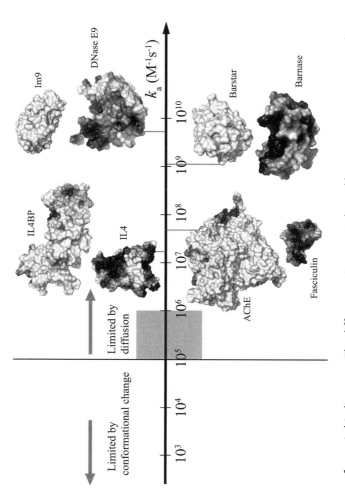

FIGURE 4.4 The spectrum of protein binding rates. The different regimes are indicated by arrows. For protein complexes in the electrostatically-enhanced regime are shown. Adapted from Alsallaq and Zhou [19].

To go beyond the basal rate k_{D0} and reach rates like 10^8 to 10^9 M^{-1} s^{-1} as observed for many protein complexes [62–69] (Figure 4.4), intermolecular forces must be present. For a force to speed up a diffusion-controlled binding, it must be present in the diffusion process that leads to the transient complex, and thus be long-ranged. Indeed, analytical results on model systems show that, when the range of the force is reduced, the resulting rate enhancement decreases drastically [57,70,71]. For protein–protein and protein–nucleic acid binding, the dominant long-range force is provided by electrostatic interactions. Rates higher than k_{D0} require favorable electrostatic interactions, which are manifested by complementary charge distributions on the two binding partners.

4.2 Brownian dynamics simulations

Many groups have used Brownian dynamics (BD) simulations to calculate the diffusion-controlled rate k_D [56,72–84] or to generate the loosely defined encounter complex [85–87]. In rate calculations one must specify a precise set of conditions, which when satisfied signifies the formation of the native complex. Specifying this set of conditions, typically implemented as an absorbing boundary in BD simulations, is equivalent to defining the transient complex. Rather than being guided by any theoretical considerations, the location of the absorbing boundary is typically proposed in an ad hoc way, and often adjusted for best agreement with experiment. Two alternative algorithms are available for obtaining k_D from statistics accumulated on BD trajectories. In one the trajectories of a protein are started from a spherical surface around the receptor [88]. In the other, the pair of binding molecules are started in the vicinity of the absorbing boundary [89].

Electrostatic interactions are accounted for by their influence on the translational and rotational Brownian motion of the binding molecules. In principle, the electrostatic force and torque on the molecules can be calculated from solving the PB equation. However, solving the PB equation on the fly during a BD simulation is prohibitively expensive. One thus has to rely on approximations, such as treating one of the proteins as a set of test charges [72] (which leads to significant errors from neglecting the low-dielectric region of the protein interior [73]) or a more elaborate effective-charge model [90]. Unfortunately, the approximations are worst when the proteins are in close proximity, precisely where electrostatic interactions are expected to have the strongest influence on k_D.

4.3 Transient-complex theory

From BD simulations [54,73,91,92] and analytical results [57,93], it was discovered that the diffusion-controlled rate can be accurately approximated as

$$k_D = k_{D0}\, e^{-\langle W_{el}\rangle^*/k_B T} \tag{7}$$

where k_{D0} is the basal rate, i.e., the rate when electrostatic interactions are turned off, and the average $\langle \cdots \rangle^*$ is over the configurational space of the transient complex. This equation resolves one of the two main obstacles to reliable prediction of binding rates, by making it possible to rigorously treat electrostatic interactions. The effect of electrostatic interactions is captured by the Boltzmann factor

$e^{-\langle W_{el}\rangle^*/k_BT}$, which can be obtained by averaging over a relatively small number of representative configurations in the transient complex. The basal rate k_{D0} still needs to be obtained through force-free BD simulations, but these simulations are inexpensive.

The remaining obstacle to reliable prediction of protein association rates is the specification of the transient complex. The ad hoc way by which the set of conditions for complex formation—which is the equivalent of the transition complex—is specified in BD simulations is noted above. The application of Eq. (7) for predicting k_D faces a similar situation. In an early application to the binding of barnase and barstar [94], the transient-complex ensemble was specified by adjusting the ranges of translation and rotation between the two proteins to match the experimental data at high ionic strength. Similarly, Miyashita et al. [95] used experimental data for the binding of cytochrome $c2$ and bacterial reaction center to locate the transient-complex ensemble in the 6-dimensional translation-rotation configurational space.

For Eq. (7) to have predictive power, the transient-complex ensemble has to be specified without reference to experiment. A solution to this challenging problem was proposed in a recent paper [15], based on analyzing the interaction energy landscape of binding proteins. The basic idea is as follows. In a complete theory, the overall binding rate k_a should not be sensitive to where the transient complex is placed. If it is placed far away from the native complex, then k_D will be large but k_c will be small. Conversely, if it is placed very close to the native complex, then k_D will be reduced but k_c will become very large. Either way, Eq. (4) is expected to give nearly the same result for k_a. However, given the considerable difficulty and uncertainty in the calculation of k_c, it is highly desirable to use k_D as a good approximation for k_a. Then there is an optimal location for placing the transient complex [96]. If it is placed too far from the native complex, then the resulting k_D would not be a useful approximation for k_a. On the other hand, placing the transient complex too close to the native complex would mean that short-range interactions and conformational rearrangement have to be dealt with in calculating k_D. The native complex sits in a deep well in the interaction energy landscape [15]. The optimal placement for the transient-complex ensemble is at the outer boundary of the native-complex energy well [15,96].

The specific procedure implementing this basic idea was based on the following observation: inside the native-complex energy well, translation and rotation are restricted, but once outside the two proteins gain significant translational and rotational freedom [15]. Thus the outer boundary of the native-complex energy well coincides with the onset of translational and rotational freedom. This onset was located by monitoring the allowed range of a relative rotation angle between the proteins as they move out of the native-complex energy well.

This structural model for the transient-complex ensemble along with Eq. (7) constitutes the transient-complex theory for predicting protein binding rates. In this theory, both of the obstacles faced by the traditional approach of BD simulations are resolved. Electrostatic interactions can be treated rigorously, and the transient complex is specified solely based on theoretical consideration.

In Eq. (7), only the electrostatic contribution to the interaction energy of the transient complex is included. The neglect of short-ranged non-electrostatic effects from the Boltzmann factor in Eq. (7) can be understood from two considerations. First, the transient-complex configurations are typically separated by at least one layer of solvent [15], therefore short-ranged forces such as hydrophobic and van der Waals interactions are relatively weak in the diffusion process leading to the transient complex. Second, as already noted in Section 4.1, compared to long-range interactions, short-range interactions, even when present within the transient complex, contribute much less to rate enhancement (i.e., k_D/k_{D0}). Including their contribution to the interaction energy in the Boltzmann factor will significantly overestimate their effect on rate enhancement. However, short-ranged interactions are essential for determining the location and size of the transient-complex ensemble in configurational space, which in turn affect the magnitude of k_{D0}. A transient-complex ensemble that is less restricted in translation and rotation will lead to a higher k_{D0}. Variation of the restriction in translation and rotation within the transient complex with solvent conditions or among different protein complexes can be viewed as a configurational entropy effect. The basal rate k_{D0} captures this entropy effect.

It has been noted that electrostatically enhanced protein binding exhibits an interesting tell-tale sign: the binding and unbinding rate constants show disparate dependences on ionic strength [96,97]. The binding rate decreases significantly with increasing ionic strength, whereas the unbinding rate is only modestly affected by ionic strength. The structural model for the transient-complex ensemble provides a nice explanation for the disparate effects of ionic strength. As the transient complex lies at the outer boundary of the interaction energy well and hence is close to the native complex, ionic strength is expected to screen electrostatic interactions in the two types of complexes to nearly the same extent. Hence the binding affinity and the binding rate are expected to have nearly the same dependence on ionic strength and the dissociation rate would be little affected by ionic strength.

The transient-complex theory has been put to a comprehensive test against experimental data [62–64,98] for the binding rates of four protein pairs (shown in Figure 4.4) and 23 of their mutants over wide ranges of ionic strength [52]. The ionic strength dependences of the binding rates for all the four protein pairs are predicted well by the theory. Moreover, the predictions for 23 mutants at various ionic strengths agree closely with experiment. In all there are 81 data points in the latter comparison, spanning four orders of magnitude in association rate. The theory thus appears to fulfill the promise of having truly predictive power. It reveals that, among the protein pairs and their mutants studied, the basal rate k_{D0} can differ by ~20-fold, but the bulk of the variations in k_D is due to the variations in $\langle W_{el} \rangle^*$, which ranges from 0 to -6 kcal/mol (the last value translates into a 10^4-fold rate enhancement).

The above comparison against experiment was based on calculating the electrostatic interaction energy from the linearized PB equation. It has been found that, when the full PB equation was used, agreement with experiment improved, albeit modestly [19]. This underscores the point that a rigorous treatment of electrostatic interactions is essential for the accuracy of calculated k_D.

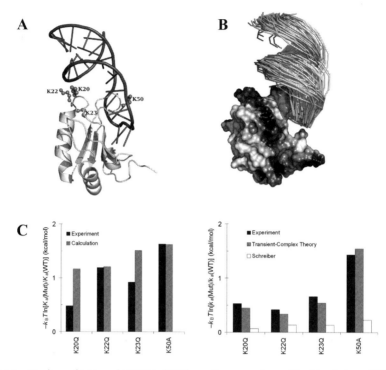

FIGURE 4.5   Binding of U1A and U1 RNA. (A) The native complex, with side chains of K20, K22, K23, and K50 of U1A shown. (B) Representative configurations in the transient complex. (C) Comparison of calculated and experimental results for the effects of mutating four lysine residues on the binding rate. Parts (A) and (B) are adapted from Qin and Zhou [71]. The left panel in (C) is taken from Qin and Zhou [18], and the right panel in (C) is taken from Qin and Zhou [71], but with the additional result label Schreiber calculated according to Selzer et al. [103].

For the binding between a protein and an RNA, the difference between the full PB equation and the linearized version is no longer modest because of the large charge density on the nucleic acid. Then use of the full PB equation becomes a necessity. The transient-complex theory has made it possible to realistically model protein–RNA binding rates for the first time [71]. In this work the binding of the spliceosomal protein U1A and its target on the U1 small nuclear RNA (Figure 4.5A) was studied. The binding and unbinding rates of this and other protein–RNA systems exhibit the disparate dependences on salt familiar to proteins [99–102], indicating that the structural model for the transient complex developed for protein–protein binding is applicable to protein–RNA binding. Representative configurations in the transient complex of the U1 system are shown in Figure 4.5B. The binding rates of the wild-type system and eight of its mutants predicted by the transient-complex theory are in close agreement with experiment [99, 101] (Figure 4.5C).

Comparison of predicted and experimental binding rates also help settle the debate between MS and vdW as the choice for the dielectric boundary in calculating the electrostatic interaction energy. The rate predictions summarized above have all been obtained by using the vdW protocol in calculating $\langle W_{el} \rangle^*$. With the MS protocol, the sign of $\langle W_{el} \rangle^*$ switches to positive (similar to what is seen on native complexes [16,17]) and now rate retardation is predicted [19]! For example, for the barnase–barstar pair, when the ionic strength is varied from 13 mM to 2000 mM, $\langle W_{el} \rangle^*$ calculated with the vdW protocol varied from -3.30 to -0.82 kcal/mol. Correspondingly, $\langle W_{el} \rangle^*$ calculated with MS protocol varied from 2.50 to 5.13 kcal/mol. For the latter results to be consistent with experiment, a basal rate in the order of 10^{10}–10^{11} $M^{-1} s^{-1}$ would be required, which clearly seems unphysical.

While the transient-complex theory is not appropriate for binding processes that are limited by large-scale conformational rearrangements, it can accommodate local conformational fluctuations. In particular, MD simulations have shown that charged side chains that eventually form cross-interface salt bridges in the native complex can form intramolecular salt bridges prior to reaching the transient complex [85]. More generally, local conformation populations in the transient complex will be different from those in the native complex. While applications of the transient-complex theory have so far assumed native conformations in the transient complex, more accurate calculations may require conformational sampling specifically within the transient complex.

4.4 Further approximation and rate enhancement by design

Based on Eq. (7), Schreiber and co-workers have made a further simplification [103–105]. Instead of using the transient-complex ensemble, they calculated $\langle W_{el} \rangle^*$ by applying an empirical function directly to the native complex. The empirical function effectively reduces the interaction energy calculated on the native complex to make it appropriate for the transient complex, and is parameterized on experimental data. Despite the approximation, the simplified approach has allowed them to design charge mutations that lead to as much as 250-fold increase in binding rate.

As an estimate for $\langle W_{el} \rangle^*$, a weaker version of the electrostatic interaction energy of the native complex seems capable of capturing general trends, but it has limitation in accounting for specific contributions of individual residues. This limitation is illustrated by the effects of mutating four lysine residues on the binding rate of U1A with U1 RNA. In the native complex (Figure 4.5A), K50 protrudes deeply into the RNA loop and lies above it, while K20, K22, and K23 lie below the loop. The four lysines have comparable separations from the RNA and their neutralizations reduce the binding free energy to similar extents (Figure 4.5C, left panel). The approach of Schreiber and co-workers would predict that the neutralizations reduce the binding rate to similar extents (Figure 4.5C, right panel). In the transient complex (Figure 4.5B), the RNA moves away from U1A, consequently K20, K22, and K23 are placed further away from the RNA. In contrast, because of its protruded position, the separation of K50 from RNA is not significantly reduced. As a result, in the transient complex the electrostatic contributions of K20, K22, and K23 are significantly reduced but that of K50 is nearly unchanged when compared to the native complex. This contrast between K20, K22, and K23 on the one hand and K50 on the other is supported by experimental results [99,101] (Figure 4.5C, right panel).

5. DYNAMICS WITHIN NATIVE COMPLEXES AND DURING COMPLEX FORMATION

The foregoing discussions make it clear that protein dynamics presents challenges for building structural models of protein complexes, makes important contributions to binding affinity, and is an integral part of the binding process. Recent experiments have presented direct evidence for the contribution of protein dynamics to binding affinity [106,107]. Changes in conformation and in dynamics upon complex formation have been studied by NMR [108–110] and by MD simulations [111]. Conformational rearrangements leading to native complexes have also been revealed by MD simulations [85,112], which as noted above may be required for more accurate calculations of binding rates within the transient-complex theory. These studies have laid the groundwork toward a comprehensive understanding of the roles of dynamics in protein–protein and protein–nucleic acid interactions.

6. SUMMARY POINTS

1. Structures of protein complexes are the basis for understanding protein interactions. Many of these structures will have to be built by docking. CAPRI provides a forum for critical assessment of docking methods. Methods making use of experimental or predicted interface information appear promising.
2. Predicting absolute binding free energy is still formidable, but there is significant progress in predicting relative binding free energy. Contributions of electrostatic interactions are sensitive to model details, in particular the choice of the boundary between the protein low dielectric and solvent high dielectric. Experimental data such as mutational effects on binding affinity are useful for selecting calculation models.
3. The wide variation of binding rates among protein complexes can be understood by considering rate-limiting conformational changes in one extreme and electrostatic rate enhancement in the opposite extreme. Current theory has shown predictive power for binding rates in the diffusion-controlled regime (i.e., those above $\sim 10^5$ M^{-1} s^{-1}).
4. Experiments and MD simulations are contributing toward a comprehensive understanding of the roles of dynamics in the various aspects of protein interactions.

ACKNOWLEDGMENTS

This work was supported in part by Grant GM058187 from the National Institutes of Health.

REFERENCES

1. Aloy, P., Bottcher, B., Ceulemans, H., Leutwein, C., Mellwig, C., Fischer, S., Gavin, A.C., Bork, P., Superti-Furga, G., Serrano, L., Russell, R.B. Structure-based assembly of protein complexes in yeast. Science 2004, 303, 2026–9.
2. Contreras-Moreira, B., Branger, P.A., Collado-Vides, J. TFmodeller: Comparative modelling of protein-DNA complexes. Bioinformatics 2007, 23, 1694–6.
3. Yi, M., Tjong, H., Zhou, H.X. Spontaneous conformational change and toxin binding in α7 nicotinic acetylcholine receptor: Insight into channel activation and inhibition. Proc. Natl. Acad. Sci. USA 2008, 105, 8280–5.
4. Tang, C., Clore, G.M. A simple and reliable approach to docking protein–protein complexes from very sparse NOE-derived intermolecular distance restraints. J. Biomol. NMR 2006, 36, 37–44.
5. Qin, S., Zhou, H.X. A holistic approach to protein docking. Proteins 2007, 69, 743–9.
6. Bhatnagar, J., Freed, J.H., Crane, B.R. Rigid body refinement of protein complexes with long-range distance restraints from pulsed dipolar ESR. Methods Enzymol. 2007, 423, 117–33.
7. Motiejunas, D., Gabdoulline, R., Wang, T., Feldman-Salit, A., Johann, T., Winn, P.J., Wade, R.C. Protein–protein docking by simulating the process of association subject to biochemical constraints. Proteins 2008, 71, 1955–69.
8. Zhou, H.X., Qin, S. Interaction-site prediction for protein complexes: A critical assessment. Bioinformatics 2007, 23, 2203–9.
9. Korkin, D., Davis, F.P., Alber, F., Luong, T., Shen, M.Y., Lucic, V., Kennedy, B.M., Sali, A. Structural modeling of protein interactions by analogy: Application to PSD-95. PLoS Comput. Biol. 2006, 2, e153.
10. Gunther, S., May, P., Hoppe, A., Frommel, C., Preissner, R. Docking without docking: ISEARCH—Prediction of interactions using known interfaces. Proteins 2007, 69, 839–44.
11. Yang, C.-Y., Wang, S.M. Recent advances in design of small-molecule ligands to target protein–protein interactions. In: Annual Reports in Computational Chemistry, vol. 2. 2006, pp. 197–219.
12. Huang, P.S., Love, J.J., Mayo, S.L. A de novo designed protein–protein interface. Protein Sci. 2007, 16, 2770–4.
13. Liu, S., Liu, S., Zhu, X., Liang, H., Cao, A., Chang, Z., Lai, L. Nonnatural protein–protein interaction-pair design by key residues grafting. Proc. Natl. Acad. Sci. USA 2007, 104, 5330–5.
14. Gilson, M.K., Zhou, H.X. Calculation of protein–ligand binding affinities. Annu. Rev. Biophys. Biomol. Struct. 2007, 36, 21–42.
15. Alsallaq, R., Zhou, H.X. Energy landscape and transition state of protein–protein association. Biophys. J. 2007, 92, 1486–502.
16. Vijayakumar, M., Dong, F., Zhou, H.X. Comparison of calculation and experiment implicates significant electrostatic contributions to the binding stability of barnase and barstar. Biophys. J. 2003, 85, 49–60.
17. Dong, F., Zhou, H.X. Electrostatic contribution to the binding stability of protein–protein complexes. Proteins 2006, 65, 87–102.
18. Qin, S., Zhou, H.X. Do electrostatic interactions destabilize protein-nucleic acid binding? Biopolymers 2007, 86, 112–8.
19. Alsallaq, R., Zhou, H.X. Electrostatic rate enhancement and transient complex of protein–protein association. Proteins 2008, 71, 320–35.
20. Swanson, J.M.J., Mongan, J., McCammon, J.A. Limitations of atom-centered dielectric functions in implicit solvation models. J. Phys. Chem. B 2005, 109, 14769–72.
21. Tan, C., Yang, L., Luo, R. How well does Poisson–Boltzmann implicit solvent agree with explicit solvent? A quantitative analysis. J. Phys. Chem. B 2006, 110, 18680–7.
22. Swanson, J.M.J., Wagoner, J.A., Baker, N.A., McCammon, J.A. Optimizing the Poisson dielectric boundary with explicit solvent forces and energies: Lessons learned with atom-centered dielectric functions. J. Chem. Theo. Comp. 2007, 3, 170–83.
23. Tjong, H., Zhou, H.X. On the dielectric boundary in Poisson–Boltzmann calculations. J. Chem. Theo. Comp. 2008, 4, 507–14.
24. del Alamo, M., Mateu, M.G. Electrostatic repulsion, compensatory mutations, and long-range non-additive effects at the dimerization interface of the HIV capsid protein. J. Mol. Biol. 2005, 345, 893–906.

25. Pal, G., Kouadio, J.L., Artis, D.R., Kossiakoff, A.A., Sidhu, S.S. Comprehensive and quantitative mapping of energy landscapes for protein–protein interactions by rapid combinatorial scanning. J. Biol. Chem. 2006, 281, 22378–85.

26. Ernst, J.A., Clubb, R.T., Zhou, H.-X., Gronenborn, A.M., Clore, G.M. Demonstration of positionally disordered water within a protein hydrophobic cavity by NMR. Science 1995, 267, 1813–7.

27. Damjanovic, A., Garcia-Moreno, B., Lattman, E.E., Garcia, A.E. Molecular dynamics study of water penetration in staphylococcal nuclease. Proteins 2005, 60, 433–49.

28. Still, A., Tempczyk, W.C., Hawley, R.C., Hendrikson, R. Semianalytical treatment of solvation for molecular mechanics and dynamics. J. Am. Chem. Soc. 1990, 112, 6127–9.

29. Lee, M.S., Feig, M., Salsbury Jr., F.R., Brooks III, C.L. New analytic approximation to the standard molecular volume definition and its application to generalized Born calculations. J. Comput. Chem. 2003, 24, 1348–56.

30. Onufriev, A., Bashford, D., Case, D.A. Exploring protein native states and large-scale conformational changes with a modified generalized Born model. Proteins 2004, 55, 383–94.

31. Tjong, H., Zhou, H.-X. GBr^6: A parameterization-free, accurate, analytical generalized Born method. J. Phys. Chem. B 2007, 111, 3055–61.

32. Guallar, V., Borrelli, K.W. A binding mechanism in protein–nucleotide interactions: Implication for U1A RNA binding. Proc. Natl. Acad. Sci. USA 2005, 102, 3954–9.

33. Ababou, A., van der Vaart, A., Gogonea, V., Merz Jr., K.M. Interaction energy decomposition in protein–protein association: A quantum mechanical study of barnase–barstar complex. Biophys. Chem. 2007, 125, 221–36.

34. Massova, I., Kollman, P.A. Computational alanine scanning to probe protein–protein interactions: A novel approach to evaluate binding free energies. J. Am. Chem. Soc. 1999, 121, 8133–43.

35. Reyes, C.M., Kollman, P.A. Structure and thermodynamics of RNA–protein binding: Using molecular dynamics and free energy analyses to calculate the free energies of binding and conformational change. J. Mol. Biol. 2000, 297, 1145–56.

36. Gohlke, H., Kiel, C., Case, D.A. Insights into protein–protein binding by binding free energy calculation and free energy decomposition for the Ras–Raf and Ras–RalGDS complexes. J. Mol. Biol. 2003, 330, 891–913.

37. Luo, C., Xu, L., Zheng, S., Luo, X., Shen, J., Jiang, H., Liu, X., Zhou, M. Computational analysis of molecular basis of 1:1 interactions of NRG-1beta wild-type and variants with ErbB3 and ErbB4. Proteins 2005, 59, 742–56.

38. Basdevant, N., Weinstein, H., Ceruso, M. Thermodynamic basis for promiscuity and selectivity in protein–protein interactions: PDZ domains, a case study. J. Am. Chem. Soc. 2006, 128, 12766–77.

39. Hou, T., Chen, K., McLaughlin, W.A., Lu, B., Wang, W. Computational analysis and prediction of the binding motif and protein interacting partners of the Abl SH3 domain. PLoS Comput. Biol. 2006, 2, e1.

40. Moreira, I.S., Fernandes, P.A., Ramos, M.J. Computational alanine scanning mutagenesis—An improved methodological approach. J. Comput. Chem. 2007, 28, 644–54.

41. Zoete, V., Michielin, O. Comparison between computational alanine scanning and per-residue binding free energy decomposition for protein–protein association using MM-GBSA: Application to the TCR-p-MHC complex. Proteins 2007, 67, 1026–47.

42. Almlof, M., Aqvist, J., Smalas, A.O., Brandsdal, B.O. Probing the effect of point mutations at protein–protein interfaces with free energy calculations. Biophys. J. 2006, 90, 433–42.

43. Zhou, Z., Bates, M., Madura, J.D. Structure modeling, ligand binding, and binding affinity calculation (LR-MM-PBSA) of human heparanase for inhibition and drug design. Proteins 2006, 65, 580–92.

44. Kormos, B.L., Benitex, Y., Baranger, A.M., Beveridge, D.L. Affinity and specificity of protein U1A-RNA complex formation based on an additive component free energy model. J. Mol. Biol. 2007, 371, 1405–19.

45. Audie, J., Scarlata, S. A novel empirical free energy function that explains and predicts protein–protein binding affinities. Biophys. Chem. 2007, 129, 198–211.

46. Clark, L.A., Boriack-Sjodin, P.A., Eldredge, J., Fitch, C., Friedman, B., Hanf, K.J., Jarpe, M., Liparoto, S.F., Li, Y., Lugovskoy, A., Miller, S., Rushe, M., Sherman, W., Simon, K., Van Vlijmen, H. Affinity enhancement of an in vivo matured therapeutic antibody using structure-based computational design. Protein Sci. 2006, 15, 949–60.

47. Song, G., Lazar, G.A., Kortemme, T., Shimaoka, M., Desjarlais, J.R., Baker, D., Springer, T.A. Rational design of intercellular adhesion molecule-1 (ICAM-1) variants for antagonizing integrin lymphocyte function-associated antigen-1-dependent adhesion. J. Biol. Chem. 2006, 281, 5042–9.
48. Sammond, D.W., Eletr, Z.M., Purbeck, C., Kimple, R.J., Siderovski, D.P., Kuhlman, B. Structure-based protocol for identifying mutations that enhance protein–protein binding affinities. J. Mol. Biol. 2007, 371, 1392–404.
49. Joachimiak, L.A., Kortemme, T., Stoddard, B.L., Baker, D. Computational design of a new hydrogen bond network and at least a 300-fold specificity switch at a protein–protein interface. J. Mol. Biol. 2006, 361, 195–208.
50. Zhou, H.-X. How do biomolecular systems speed up and regulate rates of processes? Phys. Biol. 2005, 2, R1–25.
51. Terlau, H., Shon, K.-J., Grilley, M., Stocker, M., Stuhmer, W., Baldomero, O.M. Strategy for rapid immobilization of prey by a fish-hunting marine snail. Nature 1996, 381, 148–51.
52. Alsallaq, R., Zhou, H.-X. Prediction of protein–protein association rates from a transition-state theory. Structure 2007, 15, 215–24.
53. Alsallaq, R., Zhou, H.-X. Electrostatic rate enhancement and transient complex of protein–protein association. Proteins 2007, 71, 320–35.
54. Zhou, H.-X., Wong, K.Y., Vijayakumar, M. Design of fast enzymes by optimizing interaction potential in active site. Proc. Natl. Acad. Sci. USA 1997, 94, 12372–7.
55. Tang, C., Iwahara, J., Clore, G.M. Visualization of transient encounter complexes in protein–protein association. Nature 2006, 444, 383–6.
56. Northrup, S.H., Erickson, H.P. Kinetics of protein–protein association explained by Brownian dynamics computer simulation. Proc. Natl. Acad. Sci. USA 1992, 89, 3338–42.
57. Zhou, H.-X. Enhancement of protein–protein association rate by interaction potential: Accuracy of prediction based on local Boltzmann factor. Biophys. J. 1997, 73, 2441–5.
58. Schlosshauer, M., Baker, D. Realistic protein–protein association rates from a simple diffusional model neglecting long-range interactions, free energy barriers, and landscape ruggedness. Protein Sci. 2004, 13, 1660–9.
59. Foote, J., Eisen, H.N. Kinetic and affinity limits on antibodies produced during immune responses. Proc. Natl. Acad. Sci. USA 1995, 92, 1254–6.
60. Hoffman, T.L., LaBranche, C.C., Zhang, W., Canziani, G., Robinson, J., Chaiken, I., Hoxie, J.A., Doms, R.W. Stable exposure of the coreceptor-binding site in a CD4-independent HIV-1 envelope protein. Proc. Natl. Acad. Sci. USA 1999, 96, 6359–64.
61. Wassaf, D., Kuang, G., Kopacz, K., Wu, Q.L., Nguyen, Q., Toews, M., Cosic, J., Jacques, J., Wiltshire, S., Lambert, J., Pazmany, C.C., Hogan, S., Ladner, C.R., Nixon, A.E., Sexton, D.J. High-throughput affinity ranking of antibodies using surface plasmon resonance microarrays. Anal. Biochem. 2006, 351, 241–53.
62. Wallis, R., Moore, G.K., James, R., Kleanthous, C. Protein–protein interactions in colicin E9 DNase-immunity protein complexes. 1. Diffusion-controlled association and femtomolar binding for the cognate complex. Biochemistry 1995, 34, 13743–50.
63. Schreiber, G., Fersht, A.R. Rapid, electrostatically assisted association of proteins. Nat. Struct. Biol. 1996, 3, 427–31.
64. Radic, Z., Kirchhoff, P.D., Quinn, D.M., McCammon, J.A., Taylor, P. Electrostatic influence on the kinetics of ligand binding to acetylcholinesterase. J. Biol. Chem. 1997, 272, 23265–77.
65. Shapiro, R., Ruiz-Gutierrez, M., Chen, C.-Z. Analysis of the interactions of human ribonuclease inhibitor with angiogenin and ribonuclease A by mutagenesis: Importance of inhibitor residues inside versus outside the C-terminal "hot spot". J. Mol. Biol. 2000, 302, 497–519.
66. Darling, R.J., Kuchibhotla, U., Glaesner, W., Micanovic, R., Witcher, D.R., Beals, J.M. Glycosylation of erythropoietin affects receptor binding kinetics: Role of electrostatic interactions. Biochemistry 2002, 41, 14524–31.
67. Uter, N.T., Gruic-Sovulj, I., Perona, J.J. Amino acid-dependent transfer RNA affinity in a class I aminoacyl-tRNA synthetase. J. Biol. Chem. 2005, 280, 23966–77.
68. Korennykh, A.V., Piccirilli, J.A., Correll, C.C. The electrostatic character of the ribosomal surface enables extraordinarily rapid target location by ribotoxins. Nat. Struct. Mol. Biol. 2006, 13, 436–43.
69. Johnson, R.J., McCoy, J.G., Bingman, C.A., Phillips Jr., G.N., Raines, R.T. Inhibition of human pancreatic ribonuclease by the human ribonuclease inhibitor protein. J. Mol. Biol. 2007, 368, 434–49.

70. Zhou, H.-X., Szabo, A. Enhancement of association rates by nonspecific binding to DNA and cell membranes. Phys. Rev. Lett. 2004, 93, 17810–1.

71. Qin, S., Zhou, H.X. Prediction of salt and mutational effects on the association rate of U1A protein and U1 small nuclear RNA stem/loop II. J. Phys. Chem. B 2008, 112, 5955–60.

72. Northrup, S.H., Reynolds, J.C.L., Miller, C.M., Forrest, K.J., Boles, J.O. Diffusion-controlled association rate of cytochrome c and cytochrome c peroxidase in a simple electrostatic model. J. Am. Chem. Soc. 1986, 108, 8162–70.

73. Zhou, H.-X. Brownian dynamics study of the influences of electrostatic interaction and diffusion on protein–protein association kinetics. Biophys. J. 1993, 64, 1711–26.

74. Gabdoulline, R.R., Wade, R.C. Simulation of the diffusional association of barnase and barstar. Biophys. J. 1997, 72, 1917–29.

75. Elcock, A.H., Gabdoulline, R.R., Wade, R.C., McCammon, J.A. Computer simulation of protein–protein association kinetics: Acetylcholinesterase-fasciculin. J. Mol. Biol. 1999, 291, 149–62.

76. Altobelli, G., Subramaniam, S. Kinetics of association of anti-lysozyme monoclonal antibody D44.1 and hen-egg lysozyme. Biophys. J. 2000, 79, 2954–65.

77. Fogolari, F., Ugolini, R., Molinari, H., Viglino, P., Esposito, G. Simulation of electrostatic effects in Fab-antigen complex formation. Eur. J. Biochem. 2000, 267, 4861–9.

78. Gabdoulline, R.R., Wade, R.C. Protein–protein association: Investigation of factors influencing association rates by Brownian dynamics simulations. J. Mol. Biol. 2001, 306, 1139–55.

79. De Rienzo, F., Gabdoulline, R.R., Menziani, M.C., De Benedetti, P.G., Wade, R.C. Electrostatic analysis and Brownian dynamics simulation of the association of plastocyanin and cytochrome f. Biophys. J. 2001, 81, 3090–104.

80. Zou, G., Skeel, R.D. Robust biased Brownian dynamics for rate constant calculation. Biophys. J. 2003, 85, 2147–57.

81. Ermakova, E. Lysozyme dimerization: Brownian dynamics simulation. J. Mol. Model 2005, 12, 34–41.

82. Lin, J., Beratan, D.N. Simulation of electron transfer between cytochrome c2 and the bacterial photosynthetic reaction center: Brownian dynamics analysis of the native proteins and double mutants. J. Phys. Chem. B 2005, 109, 7529–34.

83. Gross, E.L., Rosenberg, I. A Brownian dynamics study of the interaction of phormidium cytochrome f with various cyanobacterial plastocyanins. Biophys. J. 2006, 90, 366–80.

84. Blachut-Okrasinska, E., Antosiewicz, J.M. Brownian dynamics simulations of binding mRNA cap analogues to eIF4E protein. J. Phys. Chem. B 2007, 111, 13107–15.

85. Huang, X., Dong, F., Zhou, H.X. Electrostatic recognition and induced fit in the kappa-PVIIA toxin binding to Shaker potassium channel. J. Am. Chem. Soc. 2005, 127, 6836–49.

86. Spaar, A., Dammer, C., Gabdoulline, R.R., Wade, R.C., Helms, V. Diffusional encounter of barnase and barstar. Biophys. J. 2006, 90, 1913–24.

87. Forlemu, N.Y., Waingeh, V.F., Ouporov, I.V., Lowe, S.L., Thomasson, K.A. Theoretical study of interactions between muscle aldolase and F-actin: Insight into different species. Biopolymers 2007, 85, 60–71.

88. Northrup, S.H., Allison, S.A., McCammon, J.A. Brownian dynamics simulation of diffusion-influenced bimolecular reactions. J. Chem. Phys. 1984, 80, 1517–24.

89. Zhou, H.-X. Kinetics of diffusion-influenced reactions studied by Brownian dynamics. J. Phys. Chem. 1990, 94, 8794–800.

90. Gabdoulline, R.R., Wade, R.C. Effective charges for macromolecules in solvent. J. Phys. Chem. 1996, 100, 3868–78.

91. Zhou, H.-X., Briggs, J.M., McCammon, J.A. A 240-fold electrostatic rate-enhancement for acetylcholinesterase-substrate binding can be predicted by the potential within the active site. J. Am. Chem. Soc. 1996, 118, 13069–70.

92. Zhou, H.-X., Briggs, J.M., Tara, S., McCammon, J.A. Correlation between rate of enzyme-substrate diffusional encounter and average Boltzmann factor around active site. Biopolymers 1998, 45, 355–60.

93. Zhou, H.-X. Effect of interaction potentials in diffusion-influenced reactions with small reactive regions. J. Chem. Phys. 1996, 105, 7235–7.

94. Vijayakumar, M., Wong, K.Y., Schreiber, G., Fersht, A.R., Szabo, A., Zhou, H.-X. Electrostatic enhancement of diffusion-controlled protein–protein association: Comparison of theory and experiment on barnase and barstar. J. Mol. Biol. 1998, 278, 1015–24.

95. Miyashita, O., Onuchic, J.N., Okamura, M.Y. Transition state and encounter complex for fast association of cytochrome c2 with bacterial reaction center. Proc. Natl. Acad. Sci. USA 2004, 101, 16174–9.

96. Zhou, H.-X. Disparate ionic-strength dependencies of on and off rates in protein–protein association. Biopolymers 2001, 59, 427–33.

97. Zhou, H.-X. Association and dissociation kinetics of colicin E3 and immunity protein 3: Convergence of theory and experiment. Protein Sci. 2003, 12, 2379–82.

98. Shen, B.J., Hage, T., Sebald, W. Global and local determinants for the kinetics of interleukin-4/interleukin-4 receptor alpha chain interaction. A biosensor study employing recombinant interleukin-4-binding protein. Eur. J. Biochem. 1996, 40, 252–61.

99. Katsamba, P.S., Myszka, D.G., Laird-Offringa, I.A. Two functionally distinct steps mediate high affinity binding of U1A protein to U1 hairpin II RNA. J. Biol. Chem. 2001, 276, 21476–81.

100. Milev, S., Bosshard, H.R., Jelesarov, I. Enthalpic and entropic effects of salt and polyol osmolytes on site-specific protein–DNA association: The integrase Tn916-DNA complex. Biochemistry 2005, 44, 285–93.

101. Law, M.J., Linde, M.E., Chambers, E.J., Oubridge, C., Katsamba, P.S., Nilsson, L., Haworth, I.S., Laird-Offringa, I.A. The role of positively charged amino acids and electrostatic interactions in the complex of U1A protein and U1 hairpin II RNA. Nucl. Acids Res. 2006, 34, 275–85.

102. Auweter, S.D., Fasan, R., Reymond, L., Underwood, J.G., Black, D.L., Pitsch, S., Allain, F.H. Molecular basis of RNA recognition by the human alternative splicing factor Fox-1. EMBO J. 2006, 25, 163–73.

103. Selzer, T., Albeck, S., Schreiber, G. Rational design of faster associating and tighter binding protein complexes. Nat. Struct. Biol. 2000, 7, 537–41.

104. Kiel, C., Selzer, T., Shaul, Y., Schreiber, G., Herrmann, C. Electrostatically optimized Ras-binding Ral guanine dissociation stimulator mutants increase the rate of association by stabilizing the encounter complex. Proc. Natl. Acad. Sci. USA 2004, 101, 9223–8.

105. Shaul, Y., Schreiber, G. Exploring the charge space of protein–protein association: A proteomic study. Proteins 2005, 60, 341–52.

106. Horn, J.R., Kraybill, B., Petro, E.J., Coales, S.J., Morrow, J.A., Hamuro, Y., Kossiakoff, A.A. The role of protein dynamics in increasing binding affinity for an engineered protein–protein interaction established by H/D exchange mass spectrometry. Biochemistry 2006, 45, 8488–98.

107. Frederick, K.K., Marlow, M.S., Valentine, K.G., Wand, A.J. Conformational entropy in molecular recognition by proteins. Nature 2007, 448, 325–9.

108. Shajani, Z., Drobny, G., Varani, G. Binding of U1A protein changes RNA dynamics as observed by 13C NMR relaxation studies. Biochemistry 2007, 46, 5875–83.

109. Lee, D., Walsh, J.D., Yu, P., Markus, M.A., Choli-Papadopoulou, T., Schwieters, C.D., Krueger, S., Draper, D.E., Wang, Y.X. The structure of free L11 and functional dynamics of L11 in free, L11-rRNA(58 nt) binary and L11-rRNA(58 nt)-thiostrepton ternary complexes. J. Mol. Biol. 2007, 367, 1007–22.

110. Jonker, H.R., Ilin, S., Grimm, S.K., Wohnert, J., Schwalbe, H. L11 domain rearrangement upon binding to RNA and thiostrepton studied by NMR spectroscopy. Nucl. Acids Res. 2007, 35, 441–54.

111. Grunberg, R., Nilges, M., Leckner, J. Flexibility and conformational entropy in protein–protein binding. Structure 2006, 14, 683–93.

112. Bui, J.M., Radic, Z., Taylor, P., McCammon, J.A. Conformational transitions in protein–protein association: Binding of fasciculin-2 to acetylcholinesterase. Biophys. J. 2006, 90, 3280–7.

CHAPTER 5

Analyzing Protein NMR pH-Titration Curves

Jens Erik Nielsen*

Contents		

1. INTRODUCTION

The ability of certain chemical compounds to lose or gain protons has been an active area of research since the formulation the concept of pH in 1909 [1] and the appearance of the Brønsted–Lowry acid–base theory in 1923. According to Brønsted and Lowry an acid is a compound that can donate a proton, whereas a base is a compound that can accept a proton. The dissociation of a proton from an acid in solution can be modeled by a simple equilibrium constant

$$K_a = \frac{[H^+][A^-]}{[HA]},$$ (1)

* School of Biomolecular and Biomedical Science, Centre for Synthesis and Chemical Biology, UCD Conway Institute, University College Dublin, Belfield, Dublin 4, Ireland. E-mail: Jens.Nielsen@ucd.ie

Annual Reports in Computational Chemistry, Vol. 4
ISSN 1574-1400, DOI: 10.1016/S1574-1400(08)00005-4

which often is rewritten as the Henderson–Hasselbalch (HH) equation:

$$pH = pK_a + \log \frac{[A^-]}{[HA]}, \tag{2}$$

where

$$pH = -\log[H^+], \tag{3}$$

and

$$pK_a = -\log K_a. \tag{4}$$

The pK_a of a compound is thus a description of the tendency of a compound to donate its titratable proton. A plot of the fractional proton occupancy of an acid or a base versus pH is called a titration curve and has the familiar sigmoid shape typical of binding reactions.

The pK_a value of a compound is an important characteristic because it gives information on the protonation state and the charge of a compound as a function of pH. In proteins, and biomolecules in general, charges play an essential role in catalyzing reactions, determining solubility and stability, and are critical determinants of when compounds can interact with each other. The importance of pK_a values throughout last century is demonstrated by the large body of literature on the general effect of pH on biomolecules [2–4] and countless titration studies of Hen Egg White Lysozyme and other proteins (e.g. [5–18]), and pH (and thus charge) arguably constitutes the most tightly constrained and well controlled property in any biological system *in vitro* and *in vivo*. The last two decades has seen the development of methods for the accurate prediction of electrostatic properties [19,20] and pK_a values [6,21–23] from protein structures, and experimental scientists have continued to take a strong interest in pH-dependent and electrostatic phenomena of proteins with the publication of several key studies on the effect of pH on protein structure and function [24–29]. It is not the purpose of this article to review the general field of protein electrostatics, which has been reviewed extensively elsewhere [30–32], nor is it the intention to give a detailed account on the progress in the field of pK_a calculations [33–38]. Instead, I will highlight a number of studies on the analysis of NMR protein titration curves and discuss the implications and future perspectives that these studies have for our understanding of pK_a values and protein electrostatic effects.

In the following I will give an introduction to the measurement and calculation of protein residue titration curves, and continue to discuss four key papers that deal with NMR titration curve fitting: non-Henderson–Hasselbalch curve fitting [39], general equations for biomolecular titration [40], studies of electric fields in proteins [41], and studies of pH-induced conformational changes in β-lactoglobulin [42]. Collectively these studies point to interesting future directions for the theoretical and experimental study of pH-dependent effects in proteins.

1.1 Measuring protein titration curves

While titration curves can easily be measured for simple acids and bases in dilute aqueous solutions with simple titration experiments, the titration curves of individual amino acid residues in proteins are more difficult to measure. Although the overall titration curves for proteins can be measured, this rarely gives information on the titration curves of individual amino acid residue side chains since the measured overall titration curve consists of the combined titration of all the titratable groups in the protein. It is possible to gain information on the titration of individual amino acid residues by utilizing difference titration experiments; these have been performed on several proteins [15,43,44] and have yielded valuable information on the stability and activity of proteins. The high concentration of protein needed and the assumptions implicit in this method [45] limit its usefulness, and presently the preferred method for measuring titration curves of individual residues in proteins consists of recording the NMR chemical shift of individual atoms as a function of pH. Other methods such as ITC measurements [46], mass spectroscopy [47] and FT-IR spectroscopy [7] are also used.

Once titration curves have been measured using one of the above methods, they are typically fitted to Eq. (2) in the case of a single titration, or to an equation describing the pH-dependence of the chemical shift due to two or more titrations. The overall goal of experimental measurements has hitherto been to extract pK_a values from titration curves, and the information published in research publications (the pK_a values) are thus an interpretation of titration curves and not the primary experimental data.

In the case of titration curves that can be described accurately by Eq. (2), this is not a problem. However, often researchers accept less-than-perfect fits when extracting single pK_a values using Eq. (2), thus causing loss of experimental detail. In the case of titration curves displaying two or more titrations it is of particular importance to report the primary data since such titration curves can be interpreted in many different ways. This is the case for NMR titration curves, where it often is not possible to attribute a chemical shift change unambiguously to a single titrational event (see later).

The pH-dependence of the atom-specific NMR chemical shift, which is the property tracked when measuring titration curves by NMR, is influenced by any change in the chemical environment of the atom being monitored. For atoms very close (1–2 bonds away) to a titratable group, it is generally assumed that the change in the chemical shift exclusively reflects a change in the charge of the titratable group. However, if the chemical shift is measured at the backbone amide nitrogen or backbone amide hydrogen of a titratable group, the effects of other titrations can influence the chemical shift of the atom through electrostatic effects or conformational changes. In these cases it can therefore be difficult to determine if an observed titration in the chemical shift of a backbone atom of a titratable residue is due to a titrational event in that residue, due to a conformational change in the protein structure, or due to a change in the electrostatic field in the protein. Both of the latter reasons for changes in the chemical shift are ultimately due to the titration of a protein residue or group of protein residues, but in these cases the effects of a titration can be felt throughout the protein and are not limited to the atoms surrounding the titratable group. The studies on the titration curves of the active site cysteines of *E.coli* thioredoxin illustrates the difficulty of relating chemical shift changes to protonation state changes [39,48–51] since it still is unresolved if both or just one of the active site cysteines titrate around neutral pH despite a large amount of NMR titrations and other experiments on the system.

Since the analysis of NMR titration curves is not unambiguous it is important to report full titration curves in publications to allow for future re-analysis of the primary data. Fortunately titration curves have been measured for many proteins and reported in figures which allows for the extraction of the original chemical shift data using programs like g3data [52], but a large number of titration curves have been reported only as pK_a values and deposited as such in databases [53]. Experimental pK_a values are highly valuable for deriving empirical rules on the determinants of protein residue pK_a values [33,54] and for benchmarking protein pK_a calculation methods [55–58], but it is nevertheless desirable to move towards a situation where NMR titration curves are available in electronic databases. To facilitate this, we have developed a database system for protein titration curves [59]. This database allows for the upload, download and analysis of protein titration curves thus providing access to the primary experimental data produced by protein titration experiments.

1.2 Theoretical description of a titrating biomolecule

The theoretical description of the titration curves of proteins and polyprotic acids has been studied since the pioneering work of Linderstrøm-Lang [2] and Tanford [3]. In the following I will introduce the theory presently used in the calculation and analysis of protein residue titration curves.

Consider a protein with N titratable groups that each can exist in a charged and a neutral protonation state. Such a protein can occupy 2^N different protonation states at any given pH value, and each of the 2^N protonation states is associated with a free energy difference relative to an arbitrary reference state (this is normally chosen to be the fully protonated state of the protein). The pH-dependent energy difference between the reference state and a given protonation state is determined by the local environment of the titratable groups and by how they interact with each other. This is, in turn, determined by the nature and position of the titratable groups in the protein, and therefore ultimately determined by the structure (or ensemble of structures) that the protein occupies at the pH value in question.

If the relative energies for all the 2^N protonation states are known at a given pH, then the fractional charge for each titratable group (f_i) can be determined by evaluating the Boltzmann sum for the states that have a specific group in its charged state:

$$f_i = \frac{\sum_{j=1}^{2^N} \gamma_i e^{\frac{-Ex}{kT}}}{\sum_{m=1}^{2^N} e^{\frac{-Ex}{kT}}}. \tag{5}$$

Here, γ_i is a parameter that is 1 if group i is charged in state j, E_j and E_m are the energies of protonation states j and m, k is Boltzmann's constant and T is the temperature. The sum in the numerator only includes contributions from states where group i is charged.

Once the fractional charge at a given pH can be evaluated, it is straightforward to calculate the titration curves of all groups in the system by evaluating Eq. (5) at all pH values of interest. For many proteins evaluation of Eq. (5) becomes impossible due to the large number of possible protonation states, and in these cases evaluation of Eq. (5) can be substituted with Monte Carlo iterations [60] or other approximate methods [23,61] to give reliable results. The Tanford–Roxby algorithm [4] can also be used for weakly coupled systems of titratable groups, but breaks down when titratable groups interact strongly.

1.3 Describing the energy of a protonation state

It is trivial to evaluate the Boltzmann sum or use one of the approximate methods to get an accurate description of the protonation state of a protein at a specific pH once the relative energy for all protonation state are known. It has proven convenient to model the energies of the 2^N protonation states using an energy function which splits the energy into a contribution originating from the intrinsic tendency for the group to give up its proton within its local non-titrating environment, and a contribution modeling the interaction of the group with all other titratable groups. The interaction with the non-titrating environment is modeled by the so-called intrinsic pK_a value, and the interactions with the titrating environment is modeled as a sum of pair wise interaction energies. The full energy function for a particular state thus takes the form:

$$E_x = \sum_{i=1}^{N} \gamma_i \ln(10)kT(pH - pK_{a,int,i}) + \sum_{i=1}^{N}\sum_{j=1}^{N} \gamma_i \gamma_j E(i,j) \tag{6}$$

where E_x is the energy of protonation state x, $pK_{a,int,i}$ is the intrinsic pK_a of group i, $E(i,j)$ is the interaction energy between the charged forms of group i and j, γ_i and γ_j are parameters that are 1 when groups i and j are charged and zero when they are neutral.

An energy function of this type assumes that the intrinsic pK_a does not change with pH, and that the interactions with other titratable groups can be described by a sum of pair wise interaction terms. Both of these assumptions break down when the structure of a protein changes significantly with pH, but Eq. (6) nevertheless provides a convenient function for describing the energy of a protonation state; not least because all energy terms are intuitively understandable and can be evaluated from a single protein structure using a minimum of energy calculations. Despite the simplicity of Eq. (6), titration curves described by this equation can be remarkably complex. Figure 5.1 shows an example of a hypothetical system of three titratable groups that interact quite strongly with each other. It is clearly seen that the behavior of the system is anything but intuitive given the simple arrangement of groups.

Further insight into the behavior of systems of titratable groups comes from observing the population of individual protonation states. Panel B in Figure 5.1 shows the populations of the eight protonation states for the three-group system, and it is seen that the titration curves of the system are the sum of several protonation states populations. Each of the protonation state population curves are

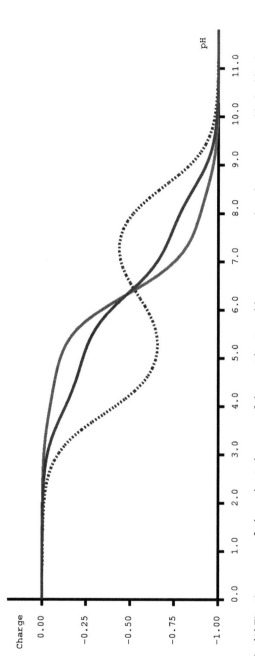

FIGURE 5.1 (A) Titration curves of a hypothetical system of three acidic titratable groups arranged so that group one (dashed line) interacts strongly with the two other groups (4 kT with group two, and 6 kT with group three), whereas the interaction between groups two (light gray) and three (dark gray) is only 2 kT. The intrinsic pK_a values of the three groups are one: 4.0, two: 5.0 and three: 4.5. Energies for each protonation state are calculated with Eq. (5).

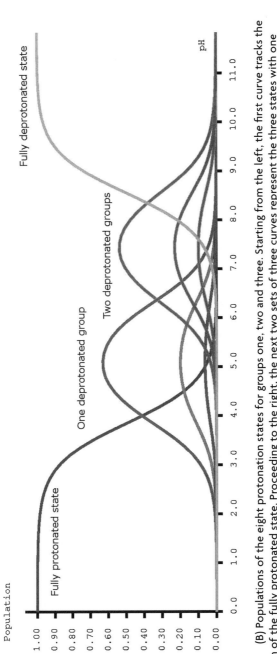
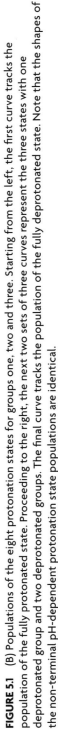

FIGURE 5.1 (B) Populations of the eight protonation states for groups one, two and three. Starting from the left, the first curve tracks the population of the fully protonated state. Proceeding to the right, the next two sets of three curves represent the three states with one deprotonated group and two deprotonated groups. The final curve tracks the population of the fully deprotonated state. Note that the shapes of the non-terminal pH-dependent protonation state populations are identical.

either sigmoid or bell-shaped. This is always the case, and non-sigmoid titration curves thus only arise when they represent the sum of several protonation state population curves.

2. FITTING PROTEIN NMR TITRATION CURVES

Classically, titration curves were fitted to gain information on the pK_a values associated with each titration. Such titration curve fitting procedures are relatively well established, mostly consist of fitting titration curves to Eq. (2) and extensions of it, and will not be covered here. Instead I will focus on extracting additional information from protein NMR pH titration curves.

2.1 Fitting non-Henderson–Hasselbalch titration curves

In the previous section it was demonstrated that protein titration curves can be calculated once we have knowledge of the intrinsic pK_a values and the pair wise interaction energies in a system. However, since we can experimentally measure titration curves, it makes sense to investigate if can we gain information on intrinsic pK_a values and pair wise interaction energies by fitting titration curves to a model composed of Eqs. (5) and (6). We can expect to gain information from such a titration curve fit if, and only if, a unique combination of the relevant parameters (the intrinsic pK_a values and pair wise interaction energies) produce the titration curves we are trying to fit. Since Henderson–Hasselbalch (HH) shaped titration curves always can be reproduced by an infinite number of parameter combinations, we are left with attempting to fit non-HH shaped titration curves.

Non-HH shaped titration curves are found when two or more groups titrate in the same pH range if they, at the same time, interact strongly [62]. Since electrostatic interaction energies decrease with $1/r$ and are much stronger in solvent inaccessible parts of a protein, we can expect to find non-HH titration curves for titratable groups that are buried. Most titratable groups in proteins are situated on the surface, but a significant fraction are situated in active sites, and these are often removed from solvent, thus producing quite strong electrostatic interactions. Titration curves have been measured only for a limited number of active site residues, and only a few of these have been found to be non-HH shaped. This could be interpreted as proof that only very few enzyme active sites have non-HH titration curves, but since titration curves have been measured primarily for small enzymes due to the size limitation of NMR experiments, and since electrostatic interaction energies are known to be stronger for larger proteins, it is quite conceivable that a large fraction of enzymes active site titratable groups display non-HH titration curves.

This is further bolstered by the fact that pK_a calculations on large enzymes routinely produce non-HH titration curves. Indeed the THEMATICS algorithm [63] for identifying active sites in protein structures has been designed around this principle, and it has been shown that strong unfavorable electrostatic interaction energies often are present in enzyme active sites [64]. It is therefore likely that a large fraction of enzyme active sites contain titratable groups that titrate in a distinctly non-HH fashion.

The work of McIntosh and co-workers on *Bacillus circulans* xylanase (BCX) [65–68] serves as one of the paradigms for the current understanding of the coupling between titration curves and pH-activity profiles. The catalytic mechanism of BCX involves a proton donor (E172) and a nucleophile (E78), and the NMR titration curves measured at atoms very close to the titratable groups of both residues are distinctly non-HH shaped. We assume that the NMR titration curves are representative of the real titration curves of the system, and we can therefore attempt to fit the E78 and E172 titration curves using Eqs. (5) and (6). Søndergaard et al. [39] implemented such a fitting routine (called Global Fitting of Titration Events—GloFTE) and were able to fit the titration curves of wild type BCX with good accuracy. The fit of these titration curves produced two intrinsic pK_a values and a value for the E78–E172 interaction energy as specified by Eq. (5). A full combinatorial scan of all realistic values for the two intrinsic pK_a values and the interaction energy confirmed that the solution found by the GloFTE fit was unique and optimal. Of the fitted parameters, the intrinsic pK_a values can be influenced by weak interactions with other titratable groups not included in the fit, and these therefore cannot be taken as a measure of the true intrinsic pK_a values of the titratable groups. Electrostatic interaction energies found by such a fit, are, however, to a very good estimation variables which uniquely describes the interaction between the two residues. Søndergaard et al. used the GloFTE algorithm on five enzyme systems and were able to extract electrostatic interaction energies for 15 residue pairs. Structural information is available for all of the systems, and it thus straightforward to calculate the effective dielectric constant for the interactions.

The ability to fit titration curves to an energy model based on Eqs. (5) and (6) also yields information on the population of each protonation state of the system and, since the catalytic groups in an enzyme active site have to be in a specific protonation state for catalysis to occur, this means that pH-activity profiles can be fitted simultaneously with the NMR titration curve. This places a further restraint on the fitting procedure and results in higher confidence levels for the extracted electrostatic interaction energies. For the majority of the enzyme systems studied by Søndergaard et al. it is indeed possible to achieve excellent agreement simultaneously for titration curves and the pH-activity profile, thus signifying that the pH-activity profile is determined by the protonation states of the active site residues and is not limited by other factors such as protein stability. It is important to fit the k_{cat}/K_m pH-activity profile simultaneously with titration curves measured for the apo form of the enzyme, while the k_{cat} pH-activity profile must be fitted simultaneously with titration curves for the holo form of the enzyme-substrate complex. This relationship is due to the pH-dependence of the Michaelis–Menten parameters and has been explored extensively [69,70].

2.1.1 Linking structure, biophysics and function
Since pairwise electrostatic interaction energies can be calculated from protein structures using Poisson–Boltzmann Equation (PBE) solvers [71,72], we can attempt to forge a unique link between protein structure and protein titration curves

by comparing calculated and fitted pair wise electrostatic interaction energies. Since we already have linked protein titration curves with pH-activity profiles, we should therefore be able to make the elusive link between protein structure, biophysics and function. Søndergaard et al. found that it was indeed possible to uniquely identify the titratable residues in the protein structure using only the fitted pair wise electrostatic interaction energies. This worked particularly well for wild type BCX and the BCX mutant N35D, whereas in other cases it was not possible to identify the correct titratable groups in the protein structure. When we cannot link the biophysics to the protein structure, then it is clear that there something is wrong. The most likely reasons are:

(1) the fitted electrostatic interaction energies are inaccurate,
(2) the calculated interaction energies are unrealistic,
(3) a change in the NMR chemical shift does not be signify a change in protonation state, or
(4) the protein occupies multiple protein conformations with different pK_a values or Eq. (6) is otherwise not able to model the energy levels of the system correctly.

Often things get interesting when they go wrong, and with NMR titration curve fitting this is indeed the case. Of the four reasons listed above, each represents a challenge to scientists studying NMR titration curves, but whereas the accuracy of the fitted interaction energies (possibility (1)) can be solved by careful statistical analysis, the three remaining possible problems are more challenging and will be discussed briefly below.

Inaccurate calculated electrostatic energies   There are many reasons that PBE-calculated electrostatic interaction energies can be incorrect. Chief among them is the infamous protein dielectric constant, which essentially is a fudge-factor that is used to scale electric fields in proteins to make them agree with experimental data. A dielectric constant is only properly defined in a macroscopic setting, and its use on a microscopic scale is problematic as pointed out by Warshel and co-workers in a number of articles [30,73]. In the study above, a protein dielectric constant of 8 was used throughout since this value has been found to work well for protein pK_a calculations [58] for a subset of proteins. However, it is entirely possible that it would be more appropriate to use a different protein dielectric constant for the proteins where we cannot relate the fitted electrostatic interaction energies to a protein structure. The problem is that we have no objective way of determining what the dielectric constant for a specific protein. Nevertheless in some cases it is clear that a physically unrealistic protein dielectric constant must be applied in order to obtain agreement between the fitted titration curves and the protein structure, and in these cases a realistic link between protein structure and biophysics therefore cannot be formed by adjusting the protein dielectric constant.

The NMR chemical shift does not model the titrational event   If it is clear that an altered protein dielectric constant cannot reconcile the protein structure with the titration curves, then it is time to examine the NMR titration curves themselves. Since the NMR chemical shift is very sensitive to small changes in the chemical environment, it is possible that a titration of a distant titratable group is mirrored in the chemical shift of another titratable group. In these cases it is very easy to mistake the 'ghost titration' originating from a distant site for a real titration of the titratable group itself.

Invalid GloFTE model assumptions   Finally, it is possible that the underlying energy function for the GloFTE method (Eq. (6)) is too simple to accurately describe the physical reality of the system. Eq. (6) operates with only a single conformational state and two protonation states for each titratable group, but it is well known that proteins often exist is several distinct conformations and furthermore that changes in pH can cause conformational changes to a protein thus invalidating the single-conformation assumption.

In a more elaborate study of protein titration, Onufriev et al. [40] convincingly showed that Eq. (6) is unable to reproduce the highly irregular titration curves for the compound diethylene-triamine-penta-acetate (DTPA) meaning that in some cases Eq. (6) simply does not describe the physics of the molecule well. In other words: for some molecules one cannot describe the energy levels of each protonation state simply by a combination of intrinsic pK_a values and pair wise interaction energies. Onufriev *et al.* developed the Decoupled Sites Representation (DSR) model for fitting and analyzing more complicated systems, and were able to fit the highly irregular titration curves of DTPA using the DSR model. The DSR model is significant because it allows us to understand the titration of any molecule in terms of a number of so-called quasi-sites. These quasi-sites do not represent a real physical titration site, but consist of linear combinations of the real titrations. This DSR model thus shows that for some compounds, their titrational behavior is better analyzed by analyzing the titratable properties of the system as a whole rather than focussing on the behavior of individual groups.

In cases where GloFTE fitting fails to agree with the protein structure via structure-based PBE calculations, we must be therefore be careful to analyze the validity of the assumptions underlying the PBE calculations and the GloFTE fitting procedure. In many cases knowledge on the flexibility of the system can lead to probable explanations that eventually can lead to the reconciliation of structure and titration curves. The following section highlights two papers that have made use of the indirect relationship between charge titration curves and NMR titration curves to study the electric field and conformational changes in proteins.

2.2 Exploiting the sensitivity of the chemical shift

The unique sensitivity of the NMR chemical shift to the chemical environment is, as illustrated above (possibility no. 3), sometimes an obstacle to understanding of the titrational properties of a system. However, the high sensitivity can also be exploited to give us a information on other characteristics of a protein.

2.2.1 The effect of the electrostatic field

An electric field (E) in a protein influences the chemical shift of all nuclei. The chemical shift of the backbone amide nitrogens is affected primarily by a polarization of the atoms in the CO-N bond, whereas the chemical shift of the backbone amid hydrogen changes due to a polarization of the N-H bond [41,74,75]. The effect on the chemical shift is maximal when the electric field is parallel to bond in question, and minimal when the electric field is perpendicular to the bond.

It is often convenient to describe the effect of the chemical shift of a particular bond as

$$\Delta_{chemshift} = A_{atom} \Delta E_{charge} \cos \theta,$$

where ΔE_{charge} is the electric field due to the removal/appearance of a single charge, θ is the angle describing the alignment of the field relative to the bond being polarized, and A_{atom} is a constant describing how much the chemical shift of the atom in question changes due to a change in the electric field. Values for A_{atom} have been calculated using quantum mechanical calculations and the value of θ can be obtained from an X-ray structure, and it is therefore possible to relate a change in chemical shift for an atom directly to a change in the electric field.

Thus, one can either predict in which atoms the titration of a certain titratable group should be observed, or, if one can ascribe the titration in an atom to a specific titratable group based on e.g. a unique pK_a value or other information, then it is possible to measure the strength of the electric field at an atom due to a specific charge. Hass et al. [41] exploit this elegantly for two titratable groups and an ion in plastocyanin, and in their analysis they determine values for A_{atom} and site-specific effective dielectric constants for the electric field emanating from each charge. Moreover, because of the angular dependence of the contribution of the electric field, the authors demonstrate that it is possible to predict the positions of the three charges in the protein structure once each chemical shift change has been attributed to a specific charge.

2.2.2 Structural changes

In a separate study Sakurai and Goto [42] use the sensitivity of the NMR chemical shift to conformational changes to characterize the pH-dependent conformational changes in β-lactoglobulin. The authors measure the pH-dependent change in chemical shift of all backbone amide nitrogen and protons, and perform a global principal component analysis (PCA) on the data. From the PCA, Sakuri and Goto find that the first three principal components account for more than 80% of the variation in the data. The authors furthermore show that the first three principal components can be represented as a pH-dependent linear combination of four basis spectra consistent with the existence of several pH-dependent conformational states of this protein [76,77]. Using the basis spectra, which are derived from the PCA, the authors are able to identify the regions of β-lactoglobulin that undergo structural changes with pH. It is found that the three transitions observed in the NMR spectra are associated with changes in the chemical shift in distinct regions of the protein, thus giving information on the effect of pH on various regions of β-lactoglobulin. Moreover, the elegant global analysis of the chemical shifts produce the same pK_a values for the structural transitions found in other studies [76,77], thus providing a convincing case for the use of PCA of NMR data for the structural interpretation of pH-dependent conformational changes.

The study of Sakuri and Goto complement many other studies on the effect of pH on protein stability and structure [3,4,78–85], but provide much more detail than conventional studies. The relatively uncomplicated analysis procedure with its excellent agreement with other methods holds great promise for the use of this method on other proteins.

2.3 Comprehensive modeling of the pH-dependence of the chemical shift

The studies of Hass et al. [41] and Sakuri and Goto [42] illustrate that the pH-dependence of the NMR chemical shift can be used as a sensitive probe to study the protein electric field and pH-dependent conformational changes in proteins. The chemical shift of a given atom is thus influenced by the electric field in the protein, the conformational state of the protein, and any changes in the shielding of an atom due to through-bond effects originating from the titration of a charge nearby. In each of the applications discussed in this article (charge titration curve measurement, measurement of the electric field, and studies of conformational changes) the pH-dependence of the NMR chemical shift has been interpreted as being dominated by a single of these components.

While for some proteins the pH-dependence of the NMR chemical shift will be dominated by either electric field (EF) changes, conformational changes or directly through-bonds by the titrational event, this cannot be expected in the general case. Instead, for most proteins the pH-dependence of the NMR chemical shift will be determined by a combination of EF changes, conformational changes and through-bonds effects. When analyzing the pH-dependence of the NMR chemical shift as in the papers discussed in this article, the computational models ought to take all three effects into account. The ever-increasing sophistication of methods for predicting chemical shifts based on protein structures [86,87] holds some promise for producing theoretical models that can decompose the effects of electric fields, conformational changes and through-bond effects on the NMR chemical shift. In particular it seems straightforward to include modeling of the effects of the electric field in these software packages.

3. CONCLUSION AND OUTLOOK

The analysis of protein titration curves can yield a wealth of information on the biophysical characteristics of proteins and give valuable insights into the energetics underlying these characteristics. Enzymatic activity, protein stability, protein conformational change and protein binding characteristics are all, to a large extent, determined by protein charge and thus protein titration curves, and the study of pH-dependent effects will therefore continue to play a central role in the study of protein biophysics both *in vitro* and *in vivo*.

Coming years will see the construction of more sophisticated methods for analyzing protein titration curves as outlined in the previous section, but there will also be a demand for delivering these methods in an easy accessible way to experimentalists. The theoretical community investigating pK_a calculations and protein electrostatics has already made significant efforts to make their technologies accessible by producing web-based interfaces to many algorithms [33,35,56,88–92] and as stand-alone clients for the prediction and analysis of titration curves are beginning to become available [39,93].

In summary, the study of protein titration curves continues to give detailed insight into the energetics of protein structures. The importance of protein electrostatics for protein structure and function combined with the interesting behavior of groups of titratable systems and the unique sensitivity of the NMR chemical shift arguably makes the study of protein titration and NMR titration curves one of the most fascinating and promising areas of research in biophysical and computational chemistry.

ACKNOWLEDGMENTS

Lawrence McIntosh, Chresten Søndergaard, Predrag Kukic, Damien Farrell and Nathan Baker are thanked for insightful discussions. This work was supported by a Science Foundation Ireland PIYR award (04/YI1/M537).

REFERENCES

1. Sørensen, S.P.L.. Biochem. Z. 1909, 21, 201.
2. Linderstrom-Lang, K. On the ionization of proteins. Compt. Rend. Trav. Lab. Carlsberg 1924, 15, 1–29.
3. Tanford, C., Kirkwood, J.G. Theory of protein titration curves. I. General equations for impenetrable spheres. J. Am. Chem. Soc. 1957, 79, 5333–9.
4. Tanford, C., Roxby, R. Interpretation of protein titration curves. Application to lysozyme. Biochemistry 1972, 11, 2193–8.
5. Kuramitsu, S., Hamaguchi, K. Analysis of the acid–base titration curve of hen lysozyme. J. Biochem. (Tokyo) 1980, 87, 1215–9.
6. Bashford, D., Karplus, M. Multiple-site titration curves of proteins: An analysis of exact and approximate methods for their calculation. J. Phys. Chem. 1991, 95, 9556–61.
7. Davoodi, J., Wakarchuk, W.W., Campbell, R.L., Carey, P.R., Surewicz, W.K. Abnormally high pKa of an active-site glutamic acid residue in Bacillus circulans xylanase. The role of electrostatic interactions. Eur. J. Biochem. 1995, 232, 839–43.
8. Oliveberg, M., Arcus, V.L., Fersht, A.R. pKA values of carboxyl groups in the native and denatured states of Barnase: The pKA values of the denatured state are on average 0.4 units lower than those of model compounds. Biochemistry 1995, 34, 9424–33.
9. Kesvatera, T., Jonsson, B., Thulin, E., Linse, S. Measurement and modelling of sequence-specific pKa values of lysine residues in calbindin D9k. J. Mol. Biol. 1996, 259, 828–39.
10. Quirk, D.J., Raines, R.T. His . . . Asp catalytic dyad of ribonuclease A: Histidine pKa values in the wild-type, D121N, and D121A enzymes. Biophys. J. 1999, 76, 1571–9.
11. Banerjee, S.K., Kregar, I., Turk, V., Rupley, J.A. Lysozyme-catalyzed reaction of the N-acetylglucosamine hexasaccharide. Dependence of rate on pH. J. Biol. Chem. 1973, 248, 4786–92.
12. Kuramitsu, S., Ikeda, K., Hamaguchi, K. Effects of ionic strength and temperature on the ionization of the catalytic groups, Asp 52 and Glu 35, in hen lysozyme. J. Biochem. (Tokyo) 1977, 82, 585–97.
13. Parsons, S.M., Raftery, M.A. Ionization behavior of the cleft carboxyls in lysozyme-substrate complexes. Biochemistry 1972, 11, 1633–8.

14. Parsons, S.M., Raftery, M.A. Ionization behavior of the catalytic carboxyls of lysozyme. Effects of temperature. Biochemistry 1972, 11, 1630–3.
15. Parsons, S.M., Raftery, M.A. Ionization behavior of the catalytic carboxyls of lysozyme. Effects of ionic strength. Biochemistry 1972, 11, 1623–9.
16. Sukumar, N., Biswal, B.K., Vijayan, M. Structures of orthorhombic lysozyme grown at basic pH and its low-humidity variant. Acta Crystallogr. D Biol. Crystallogr. 1999, 55(Pt 4), 934–7.
17. Takahashi, T., Nakamura, H., Wada, A. Electrostatic forces in two lysozymes: Calculations and measurements of histidine pKa values. Biopolymers 1992, 32, 897–909.
18. Yang, Y., Hamaguchi, K. Hydrolysis of 4-methylumbelliferyl N-acetyl-chitotetraoside catalyzed by hen lysozyme. J. Biochem. (Tokyo) 1980, 88, 829–36.
19. Klapper, I., Hagstrom, R., Fine, R., Sharp, K., Honig, B. Focusing of electric fields in the active site of Cu-Zn superoxide dismutase: Effects of ionic strength and amino-acid modification. Proteins 1986, 1, 47–59.
20. Nielsen, J.E., Andersen, K.V., Honig, B., Hooft, R.W., Klebe, G., Vriend, G., Wade, R.C. Improving macromolecular electrostatics calculations. Protein Eng. 1999, 12, 657–62.
21. Bashford, D., Karplus, M. pKa's of ionizable groups in proteins: Atomic detail from a continuum electrostatic model. Biochemistry 1990, 29, 10219–25.
22. Warshel, A. Calculations of enzymatic reactions: Calculations of pKa, proton transfer reactions, and general acid catalysis reactions in enzymes. Biochemistry 1981, 20, 3167–77.
23. Yang, A.S., Gunner, M.R., Sampogna, R., Sharp, K., Honig, B. On the calculation of pKa's in proteins. Proteins 1993, 15, 252–65.
24. Lindman, S., Linse, S., Mulder, F.A., Andre, I. Electrostatic contributions to residue-specific protonation equilibria and proton binding capacitance for a small protein. Biochemistry 2006, 45, 13993–4002.
25. Schubert, M., Poon, D.K., Wicki, J., Tarling, C.A., Kwan, E.M., Nielsen, J.E., Withers, S.G., McIntosh, L.P. Probing electrostatic interactions along the reaction pathway of a glycoside hydrolase: Histidine characterization by NMR spectroscopy. Biochemistry 2007, 46, 7383–95.
26. Fitch, C.A., Karp, D.A., Lee, K.K., Stites, W.E., Lattman, E.E., Garcia-Moreno, E.B. Experimental pK(a) values of buried residues: Analysis with continuum methods and role of water penetration. Biophys. J. 2002, 82, 3289–304.
27. Schwehm, J.M., Fitch, C.A., Dang, B.N., Garcia-Moreno, E.B., Stites, W.E. Changes in stability upon charge reversal and neutralization substitution in staphylococcal nuclease are dominated by favorable electrostatic effects. Biochemistry 2003, 42, 1118–28.
28. Andre, I., Linse, S., Mulder, F.A. Residue-specific pKa determination of lysine and arginine side chains by indirect 15N and 13C NMR spectroscopy: Application to apo calmodulin. J. Am. Chem. Soc. 2007, 129, 15805–13.
29. Lindman, S., Linse, S., Mulder, F.A., Andre, I. pK(a) values for side-chain carboxyl groups of a PGB1 variant explain salt and pH-dependent stability. Biophys. J. 2007, 92, 257–66.
30. Warshel, A., Sharma, P.K., Kato, M., Parson, W.W. Modeling electrostatic effects in proteins. Biochim. Biophys. Acta 2006, 1764, 1647–76.
31. Sharp, K., Honig, B. Electrostatic interactions in macromolecules: Theory and applications. Annu. Rev. Biophys. Chem. 1990, 19, 301–32.
32. Perutz, M.F. Electrostatic effects in proteins. Science 1978, 201, 1187–91.
33. Li, H., Robertson, A.D., Jensen, J.H. Very fast empirical prediction and rationalization of protein pK(a) values. Proteins 2005, 61, 704–21.
34. Alexov, E. Role of the protein side-chain fluctuations on the strength of pairwise electrostatic interactions: Comparing experimental with computed pK(a)s. Proteins 2003, 50, 94–103.
35. Gordon, J.C., Myers, J.B., Folta, T., Shoja, V., Heath, L.S., Onufriev, A. H++: A server for estimating pKa's and adding missing hydrogens to macromolecules. Nucleic Acids Res. 2005, 33, W368–71.
36. Antosiewicz, J., McCammon, J.A., Gilson, M.K. Prediction of pH-dependent properties of proteins. J. Mol. Biol. 1994, 238, 415–36.
37. Walczak, A.M., Antosiewicz, J. Langevin dynamics of proteins at constant pH. Physical Rev. E 2002, 66, 1–8.
38. Mongan, J., Case, D.A., McCammon, J.A. Constant pH molecular dynamics in Generalized Born implicit solvent. J. Comput. Chem. 2004, 25, 2038–48.

39. Søndergaard, C.R., McIntosh, L.P., Pollastri, G., Nielsen, J.E. Determination of electrostatic interaction energies and protonation state populations in enzyme active sites by global fits of NMR-titration data and pH-activity profiles. J. Mol. Biol. 2008, 376, 269–87.
40. Onufriev, A., Case, D.A., Ullmann, G.M. A novel view of pH titration in biomolecules. Biochemistry 2001, 40, 3413–9.
41. Hass, M.A., Jensen, M.R., Led, J.J. Probing electric fields in proteins in solution by NMR spectroscopy. Proteins 2008, 72(1), 333–43.
42. Sakurai, K., Goto, Y. Principal component analysis of the pH-dependent conformational transitions of bovine beta-lactoglobulin monitored by heteronuclear NMR. Proc. Natl. Acad. Sci. USA 2007, 104, 15346–51.
43. Lewis, S.D., Johnson, F.A., Shafer, J.A. Potentiometric determination of ionizations at the active site of papain. Biochemistry 1976, 15, 5009–17.
44. Fukae, K., Kuramitsu, S., Hamaguchi, K. Binding of N-acetyl-chitotriose to Asp 52-esterified hen lysozyme. J. Biochem. 1979, 85, 141–7.
45. Migliorini, M., Creighton, D.J. Active-site ionizations of papain. An evaluation of the potentiometric difference titration method. Eur. J. Biochem. 1986, 156, 189–92.
46. Tolbert, B.S., Tajc, S.G., Webb, H., Snyder, J., Nielsen, J.E., Miller, B.L., Basavappa, R. The active site cysteine of ubiquitin-conjugating enzymes has a significantly elevated pK(a): Functional implications. Biochemistry 2005, 44, 16385–91.
47. Nelson, K.J., Day, A.E., Zeng, B.B., King, S.B., Poole, L.B. Isotope-coded, iodoacetamide-based reagent to determine individual cysteine pK(a) values by matrix-assisted laser desorption/ionization time-of-flight mass spectrometry. Anal. Biochem. 2008, 375, 187–95.
48. Chivers, P.T., Prehoda, K.E., Volkman, B.F., Kim, B.M., Markley, J.L., Raines, R.T. Microscopic pKa values of Escherichia coli thioredoxin. Biochemistry 1997, 36, 14985–91.
49. Wilson, N.A., Barbar, E., Fuchs, J.A., Woodward, C. Aspartic acid 26 in reduced Escherichia coli thioredoxin has a pKa > 9. Biochemistry 1995, 34, 8931–9.
50. Jeng, M.F., Holmgren, A., Dyson, H.J. Proton sharing between cysteine thiols in Escherichia coli thioredoxin: Implications for the mechanism of protein disulfide reduction. Biochemistry 1995, 34, 10101–5.
51. Jeng, M.F., Dyson, H.J. Direct measurement of the aspartic acid 26 pKa for reduced Escherichia coli thioredoxin by 13C NMR. Biochemistry 1996, 35, 1–6.
52. Frantz, J. G3Data. http://www.frantz.fi/software/g3data.php, 2007.
53. Toseland, C.P., McSparron, H., Davies, M.N., Flower, D.R. PPD v1.0—An integrated, web-accessible database of experimentally determined protein pKa values. Nucleic Acids Res. 2006, 34, D199–203.
54. Forsyth, W.R., Antosiewicz, J.M., Robertson, A.D. Empirical relationships between protein structure and carboxyl pKa values in proteins. Proteins 2002, 48, 388–403.
55. Davies, M.N., Toseland, C.P., Moss, D.S., Flower, D.R. Benchmarking pK(a) prediction. BMC Biochem. 2006, 7, 18.
56. Alexov, E.G., Gunner, M.R. Incorporating protein conformational flexibility into the calculation of pH-dependent protein properties. Biophys. J. 1997, 72, 2075–93.
57. Mehler, E.L., Guarnieri, F. A self-consistent, microenvironment modulated screened coulomb potential approximation to calculate pH-dependent electrostatic effects in proteins. Biophys. J. 1999, 77, 3–22.
58. Nielsen, J.E., Vriend, G. Optimizing the hydrogen-bond network in Poisson–Boltzmann equation-based pK(a) calculations. Proteins 2001, 43, 403–12.
59. Farrell, D., Sa-Miranda, E., Georgi, N., Webb, H., Nielsen, J.E. Titration_DB: A database of protein residue titration curves. http://peat.ucd.ie/titration_db, in preparation, 2008.
60. Beroza, P., Fredkin, D.R., Okamura, M.Y., Feher, G. Protonation of interacting residues in a protein by a Monte Carlo method: Application to lysozyme and the photosynthetic reaction center of Rhodobacter sphaeroides. Proc. Natl. Acad. Sci. USA 1991, 88, 5804–8.
61. Myers, J., Grothaus, G., Narayanan, S., Onufriev, A. A simple clustering algorithm can be accurate enough for use in calculations of pK's in macromolecules. Proteins 2006, 63, 928–38.
62. Klingen, A.R., Bombarda, E., Ullmann, G.M. Theoretical investigation of the behavior of titratable groups in proteins. Photochem. Photobiol. Sci. 2006, 5, 588–96.
63. Ondrechen, M.J., Clifton, J.G., Ringe, D. THEMATICS: A simple computational predictor of enzyme function from structure. Proc. Natl. Acad. Sci. USA 2001, 98, 12473–8.

64. Elcock, A.H. Prediction of functionally important residues based solely on the computed energetics of protein structure. J. Mol. Biol. 2001, 312, 885–96.

65. McIntosh, L.P., Hand, G., Johnson, P.E., Joshi, M.D., Korner, M., Plesniak, L.A., Ziser, L., Wakarchuk, W.W., Withers, S.G. The pKa of the general acid/base carboxyl group of a glycosidase cycles during catalysis: A 13C-NMR study of bacillus circulans xylanase. Biochemistry 1996, 35, 9958–66.

66. Joshi, M.D., Hedberg, A., McIntosh, L.P. Complete measurement of the pKa values of the carboxyl and imidazole groups in Bacillus circulans xylanase. Protein Sci. 1997, 6, 2667–70.

67. Joshi, M.D., Sidhu, G., Pot, I., Brayer, G.D., Withers, S.G., McIntosh, L.P. Hydrogen bonding and catalysis: A novel explanation for how a single amino acid substitution can change the pH optimum of a glycosidase. J. Mol. Biol. 2000, 299, 255–79.

68. Joshi, M.D., Sidhu, G., Nielsen, J.E., Brayer, G.D., Withers, S.G., McIntosh, L.P. Dissecting the electrostatic interactions and pH-dependent activity of a family 11 glycosidase. Biochemistry 2001, 40, 10115–39.

69. Fersht, A.R. Structure and Mechanism in Protein Science: A Guide to Enzyme Catalysis and Protein Folding. New York: Freeman; 1999.

70. Kyte, J. Mechanism in Protein Chemistry. New York/London: Garland; 1995.

71. Nicholls, A., Honig, B. A rapid finite difference algorithm, utilizing successive over-relaxation to solve the Poisson–Boltzmann equation. J. Comp. Chem. 1991, 12, 435–45.

72. Baker, N.A., Sept, D., Joseph, S., Holst, M.J., McCammon, J.A. Electrostatics of nanosystems: Application to microtubules and the ribosome. Proc. Natl. Acad. Sci. USA 2001, 98, 10037–41.

73. Schutz, C.N., Warshel, A. What are the dielectric "constants" of proteins and how to validate electrostatic models? Proteins 2001, 44, 400–17.

74. LeMaster, D.M. Structural determinants of the catalytic reactivity of the buried cysteine of Escherichia coli thioredoxin. Biochemistry 1996, 35, 14876–81.

75. Grayson, M., Raynes, W. Electric field effects on the shielding of protons in C-H bonds. Magnetic Resonance Chem. 1995, 33, 138–43.

76. Sawyer, L., Kontopidis, G. The core lipocalin, bovine beta-lactoglobulin. Biochim. Biophys. Acta 2000, 1482, 136–48.

77. Taulier, N., Chalikian, T.V. Characterization of pH-induced transitions of beta-lactoglobulin: Ultrasonic, densimetric, and spectroscopic studies. J. Mol. Biol. 2001, 314, 873–89.

78. Pace, C.N., Laurents, D.V., Thomson, J.A. pH dependence of the urea and guanidine hydrochloride denaturation of ribonuclease A and ribonuclease T1. Biochemistry 1990, 29, 2564–72.

79. Tanford, C. Protein denaturation. Part C. Theoretical models for the mechanism of denaturation. Adv. Protein Chem. 1970, 25, 1–95.

80. Yang, A.S., Honig, B. On the pH dependence of protein stability. J. Mol. Biol. 1993, 231, 459–74.

81. Alexov, E. Numerical calculations of the pH of maximal protein stability. The effect of the sequence composition and three-dimensional structure. Eur. J. Biochem. 2004, 271, 173–85.

82. Lambeir, A.M., Backmann, J., Ruiz-Sanz, J., Filimonov, V., Nielsen, J.E., Kursula, I., Norledge, B.V., Wierenga, R.K. The ionization of a buried glutamic acid is thermodynamically linked to the stability of Leishmania mexicana triose phosphate isomerase. Eur. J. Biochem. 2000, 267, 2516–24.

83. Tollinger, M., Crowhurst, K.A., Kay, L.E., Forman-Kay, J.D. Site-specific contributions to the pH dependence of protein stability. Proc. Natl. Acad. Sci. USA 2003, 100, 4545–50.

84. Pace, C.N., Alston, R.W., Shaw, K.L. Charge-charge interactions influence the denatured state ensemble and contribute to protein stability. Protein Sci. 2000, 9, 1395–8.

85. Elcock, A.H. Realistic modeling of the denatured states of proteins allows accurate calculations of the pH dependence of protein stability. J. Mol. Biol. 1999, 294, 1051–62.

86. Xu, X.P., Case, D.A. Probing multiple effects on 15N, 13C alpha, 13C beta, and 13C' chemical shifts in peptides using density functional theory. Biopolymers 2002, 65, 408–23.

87. Neal, S., Nip, A.M., Zhang, H., Wishart, D.S. Rapid and accurate calculation of protein 1H, 13C and 15N chemical shifts. J. Biomol. NMR 2003, 26, 215–40.

88. Dolinsky, T.J., Nielsen, J.E., McCammon, J.A., Baker, N.A. PDB2PQR: An automated pipeline for the setup of Poisson–Boltzmann electrostatics calculations. Nucleic Acids Res. 2004, 32, W665–7.

89. Dolinsky, T.J., Czodrowski, P., Li, H., Nielsen, J.E., Jensen, J.H., Klebe, G., Baker, N.A. PDB2PQR: Expanding and upgrading automated preparation of biomolecular structures for molecular simulations. Nucleic Acids Res. 2007, 35, W522–5.

90. Tynan-Connolly, B., Nielsen, J.E. pKD: Re-designing protein pKa values. Nucleic Acids Res. 2006, 34, W48–51.
91. Tynan-Connolly, B.M., Nielsen, J.E. Re-Designing protein pKa values. Protein Sci. 2007, 16, 239–49.
92. Georgescu, R.E., Alexov, E.G., Gunner, M.R. Combining conformational flexibility and continuum electrostatics for calculating pK(a)s in proteins. Biophys. J. 2002, 83, 1731–48.
93. Nielsen, J.E. Analysing the pH-dependent properties of proteins using pKa calculations. J. Mol. Graph. 2006, 25, 691–9.

CHAPTER 6

Implicit Solvent Simulations of Biomolecules in Cellular Environments

Michael Feig*, Seiichiro Tanizaki**, and **Maryam Sayadi*****

Contents

1. INTRODUCTION

The application of implicit solvent in simulations of biological macromolecules is an attractive choice over fully explicit solvent environments [1]. The primary advantage is a lower computational cost due to the reduced system size with implicit solvent, while reduced or absent solvent friction and instant solvent relaxation can further accelerate conformational sampling [2,3]. The mean-field nature of implicit solvent also results in larger system energy fluctuations which are advantageous in enhanced sampling methods and in particular in temperature replica exchange

*   Department of Biochemistry and Molecular Biology, Michigan State University, East Lansing, MI 48824, USA
     E-mail: feig@msu.edu (M. Feig)
**   Department of Chemistry and Biochemistry, University of Texas at Arlington, Arlington, TX 76019, USA
*** Department of Chemistry, Michigan State University, East Lansing, MI 48824, USA

Annual Reports in Computational Chemistry, Vol. 4
ISSN 1574-1400, DOI: 10.1016/S1574-1400(08)00006-6

simulations [4], because a smaller number of replicas can provide sufficient sampling overlap for a given range of temperatures

Numerous simulations with implicit solvent have been reported over the last decades with increasing levels of realism [5]. Many of these simulations involve peptides and small proteins over time scales that could not be reached easily with explicit solvent representations [6–8]. There are many fewer cases where implicit solvent simulations were used to study larger biomolecules and biomolecular complexes [9,10]. Such simulations have remained rare for the following reasons:

(1) The computational advantage of implicit solvent diminishes at large system sizes when the ratio of solvent vs. solute atoms decreases.
(2) Because implicit solvent simulations are typically open, non-periodic systems, electrostatic interactions are calculated according to standard cutoff schemes that do not scale as well to large system sizes as the more efficient Ewald summation method [11,12] that is routinely used in explicit solvent simulations.
(3) Solvation free energies are often estimated less accurately for large solutes because the approximate methods that are typically used have difficulties with internal solute cavities [8] and with solute surface charges that are very far from the center of the solute [13].

As a consequence, the structural integrity of larger biomolecules may not be maintained as well with implicit solvent [14].

While the implicit modeling of dilute aqueous solvent is well established, the realistic modeling of cellular environments presents additional challenges that are just beginning to be addressed. Cellular environments are densely crowded with macromolecules and co-solvents [15]. Crowding results in steric hindrance [16] but also modulates the properties of the remaining aqueous solvent [17]. Furthermore, many biological processes take place in or near phospholipid bilayer membranes where physicochemical properties differ drastically from aqueous solvent. Implicit solvent models of cellular environments need to consider the heterogeneous nature of biological systems on the molecular level while providing a mean-field model that does not require an explicit representation of solvent or co-solvent. In general, this can be achieved through spatially varying continuum models as the basis for calculating the energetic effects of a given environment.

2. THEORY

The mean-field effect of the environment can be included in biomolecular simulations simply by adding an expression for the solvation free energy of an instantaneous solute conformation to a given molecular mechanics force field [1]. Such an implicit solvent potential addresses the thermodynamic component of solute–solvent interactions. Kinetic and hydrodynamic properties may be reintroduced through the use of Langevin dynamics where coupling with a temperature bath is implemented through stochastic collisions and solvent friction [2,3,18].

Solvation free energies can be calculated with a number of empirical formalisms [19–21]. A more physical approach involves the decomposition of the solvation free energy into electrostatic and non-polar components.

The dominant electrostatic component can be obtained according to the linearized Poisson–Boltzmann equation [22] (Eq. (1)). Poisson theory describes the solute–solvent system as a two-dielectric continuum where a high dielectric environment surrounds a low-dielectric solute cavity with embedded partial atomic charges according to a given force field:

$$\nabla \cdot \left[\varepsilon(\mathbf{r}) \nabla \phi(\mathbf{r}) \right] - \kappa^2(\mathbf{r}) \phi(\mathbf{r}) = -4\pi \rho(\mathbf{r}) \tag{1}$$

where $\rho(r)$ is the explicit charge distribution, $\varepsilon(r)$ is the dielectric constant, and $\kappa(r)$ is the modified Debye–Hückel screening factor which captures interactions with free ions in the environment. In the case of low ionic concentrations, the second term on the left side may be neglected. The result is the simplified Poisson equation. Only static dielectric constants are considered here because the dielectric response is essentially constant over the frequency range corresponding to conformational fluctuations in biomolecules.

The Poisson equation can be solved for the electrostatic potential using finite difference techniques [23,24]. Such calculations are straightforward but relatively expensive [25]. There has been recent progress in improving the accuracy of solvation free energies from Poisson theory [26], but the computational cost for solving the Poisson equation directly at each simulation step in a molecular dynamics simulation has remained prohibitive.

A computationally attractive alternative to a direct solution of Eq. (1) is the popular Generalized Born (GB) formalism [5,27] given in Eq. (2) which empirically approximates electrostatic solvation free energies from Poisson theory:

$$\Delta G_{\text{elec}} = -\frac{1}{2}\left(1 - \frac{1}{\varepsilon}\right) \sum_{i,j} \frac{q_i q_j}{\sqrt{r_{ij}^2 + \alpha_i \alpha_j e^{-r_{ij}^2/F\alpha_i\alpha_j}}} \tag{2}$$

where q_i are partial atomic charges of the solute from a given force field, r_{ij} are pairwise atomic distances, and F is an adjustable parameter. The key to a successful implementation of the GB formalism is an efficient and accurate calculation of the generalized Born radii α_i as described in more detail in Chapter 7 of this volume.

The non-polar contribution to the free energy of solvation includes van der Waals solute-solvent interactions and the cost of cavity formation. The cost of cavity formation can be modeled effectively with a term that is simply proportional to the solvent accessible surface area or the solvent excluded volume [28–30]. Often, van der Waals interactions are not considered separately, but new formalisms have been suggested to include such contributions effectively [31]. Recently, it was also recognized that the length-scale dependence of non-polar interactions may further complicate the development of accurate implicit non-polar models [32]. It is clear that further studies will be needed to identify the optimal implicit non-polar formalism.

Cellular environments can be modeled as a first approximation as aqueous solvent, corresponding to an implicit solvent model with $\varepsilon = 80$. The high concentration of biomolecules and co-solvents inside cells generally attenuates the dielectric response of the remaining water molecules [17]. In a mean field approach, this corresponds to a uniformly lowered dielectric constant for the environment. In addition, molecular crowding imposes steric constraints that disfavor more extended conformations of biomolecules. This may be modeled simply as an increased cost of cavity formation.

Heterogeneity of cellular environments can be introduced by allowing the environmental dielectric constant and cost of cavity formation to vary spatially. An example where this idea has been applied successfully is the implicit modeling of biological membranes composed of phospholipid bilayers. The lipid tails in the membrane interior present a highly hydrophobic environment with a dielectric constant between 1 and 2 [33]. The environmental polarizability increases to intermediate values near the glycerol linkage before reaching dielectric constants comparable to aqueous solvent in the head group region [33]. At the same time, the cost of cavity formation decreases from a relatively large value in aqueous solvent to nearly zero in the center of the membrane where the lipid tails are easily moved to accommodate a given solute.

Lowered and spatially varying dielectric constants are readily applied in Poisson theory through a suitable definition of $\varepsilon(r)$ in Eq. (1). However, it is less obvious how to modify Eq. (2) to implement such models within the GB formalism. In fact, there are two separate issues that have to be addressed:

(1) How applicable is the GB formalism in the case of (homogeneous) low dielectric environments?
(2) How can dielectric heterogeneity be introduced?

It has been recognized previously that changing ε in Eq. (2) is not sufficient to accurately reflect solvation free energies in low dielectric environments [34,35]. The generalized Born radii α_i also have to be calculated as a function of ε [35]. This is evidenced by the higher-order terms in the analytical expression for the electrostatic solvation free energy for a single charge at distance r from the center of a spherical cavity of radius R according to Kirkwood [36]:

$$\Delta G_{elec} = -\frac{q^2}{2}\left(1-\frac{1}{\varepsilon}\right)\left(\frac{1}{R}+\frac{2\varepsilon}{2\varepsilon+1}\frac{r^2}{R^3}+\frac{3\varepsilon}{3\varepsilon+2}\frac{r^4}{R^5}+\cdots\right) \tag{3}$$

The first term in the series expansion in Eq. (3) corresponds to the Coulomb field approximation that is used in GB formalisms to calculate GB radii [37]. Born radii calculated solely based on the Coulomb field approximation are therefore not sensitive to the dielectric environment (as they should be). However, recent GB implementations that include an additional higher-order correction term [13,38] allow the ε-dependent calculation of GB radii and consequently an accurate reproduction of solvation energies over the entire range of dielectric constants [35].

To allow for heterogeneous dielectric environments a local dielectric constant that varies at each charge site can be introduced into the standard GB formalism:

$$\Delta G_{elec} = -\frac{1}{2}\sum_{i,j}\left(1-\frac{1}{(\varepsilon_i+\varepsilon_j)/2}\right)\frac{q_iq_j}{\sqrt{r_{ij}^2+\alpha_i(\varepsilon_i)\alpha_j(\varepsilon_j)e^{-r_{ij}^2/F\alpha_i(\varepsilon_i)\alpha_j(\varepsilon_j)}}}. \tag{4}$$

In the membrane bilayer example, ε may simply vary with the distance from the membrane center so that $\varepsilon_i = \varepsilon(z_i)$. This spatially variable dielectric constant represents the effective dielectric response at a given charge site that is due to the entire heterogeneous environment. In the membrane example, this means that the effective dielectric constant for a solute atom embedded in the hydrophobic region is not simply equal to the dielectric constant of the hydrophobic lipid tails but also depends on the higher dielectric regions that are at a distance. The global effect of a heterogeneous dielectric environment can be captured by solving the Poisson equation for a charged probe sphere at different sites and back-calculating the corresponding effective dielectric constant from the solvation free energy according to the inverted Born equation:

$$\varepsilon_{\text{eff}} = \frac{1}{1 + 2\Delta G_{PB}a/q^2} \tag{5}$$

where a is the radius of the probe sphere and q is its charge.

The formalism outlined so far allows for the implicit modeling of complex cellular environments as long as they can be approximated through spatially varying dielectric regions. This spatial coarse-graining may be appropriate for many applications, but neglects further variations in the environment over time. Coming back to the membrane example, phospholipid bilayers are not static entities but rather fluctuate both in terms of thickness and height [39]. A temporal variation of the environment could be realized by allowing the effective dielectric function to vary with time either driven by an external function or based on interactions with a solute interacting with the membrane system.

3. APPLICATIONS AND CHALLENGES

3.1 Simulations of biomolecules with implicit solvent

As implicit solvent models have become more realistic, the focus has shifted from method development and validation to applications. Early applications have applied implicit solvent models as scoring functions [40] and in particular for the prediction of binding affinities through the popular MMPB/SA or MMGB/SA approach where conformational ensembles from explicit solvent are rescored with an implicit solvent function [41]. More recently, implicit solvent models have also found wider application in molecular dynamics simulations. Implicit solvent is generally attractive to study the dynamics of peptides and proteins over long time scales [7]. The conformational sampling of peptides in implicit solvent has been studied with conventional, constant-temperature molecular dynamics simulations [42–44] and with multicanonical sampling [45]. Especially popular has been the use of implicit solvent to study the folding of hairpins [46–51], helices [49,52], and small proteins such as the WW domain [53] and protein A [54] as well as the unfolding of transthyretin [55]. Most of these calculations have also employed enhanced sampling methods, in particular temperature replica exchange simulations, for which implicit solvent is well suited due to the reduced degrees of freedom of the entire system.

A recent application of implicit solvent in simulations of larger systems has been reported for the nucleosome core particle [10], for chemotaxis Y protein [56], in a study of unfolded states of apomyoglobin [57], and in efforts to model the structure and dynamics of steroidogenic acute regulatory protein-related lipid transfer domains [58]. Most implicit solvent simulations that have been reported involve peptide or protein solutes, but success with implicit solvent simulations of nucleic acids has also been reported [59,60].

As mentioned above, the use of implicit solvent in conventional simulations of larger biomolecules is often not as effective because of technical and efficiency issues [61]. However, implicit solvent may be a good choice in biased or targeted molecular dynamics simulations where the lack of explicit solvent eliminates the need for solvent re-equilibration during the course of a simulation, especially when large conformational changes are involved. In exploratory simulations of yeast RNA polymerase II that examined a putative DNA/RNA translocation mechanism the use of implicit solvent allowed the entire translocation step to be simulated in a semi-realistic fashion over tens to hundreds of picoseconds [62]. This would not have been possible with explicit solvent. Similarly, ligand unbinding simulations of odorant-binding proteins took advantage of implicit solvent to eliminate the need to simulate solvent reorganization during the binding process [63].

An exciting development made possible by implicit solvent is the implementation of constant pH simulations where the ionization states of ionizable residues are allowed to vary dynamically in response to interactions with the surrounding residues and the environment [64,65]. Although the methodology was introduced only recently, interesting first applications have already been reported [66,67].

Finally, implicit solvent has been found useful in the refinement of protein structures based on experimental constraints [68] and without restraints from approximate homology models [69] when combined with replica exchange methodology.

Despite the increasing number of successes with implicit solvent simulations, there are still concerns that the use of implicit solvent systematically biases conformational sampling. In particular, studies have suggested increased sampling of helical structures [70] and an overstabilization of salt bridges [52,56,71]. These problems may be linked to deficiencies with particular implicit solvent models or they may reflect uncertainty in defining the dielectric surface based on default atomic radii, usually taken from the van der Waals radii of the underlying force field. Consequently, modified atomic radii have been found to improve the agreement between implicit solvent simulations and experimental data and/or explicit solvent simulations [49,59,71].

3.2 Ensembles, solvent friction, and conformational sampling

Simulations of biomolecules are typically carried out in NVT (constant number of particles, volume, and temperature) or NPT (constant pressure instead of volume) ensembles to reflect the conditions of biological environments. While Newtonian mechanics for a fixed simulation box with a solute and explicit solvent results in sampling of an NVE ensemble (constant energy instead of temperature), NVT ensembles can be achieved by coupling the simulated system to a temperature bath. A number of different coupling schemes have been proposed [72], all of which either rescale or reassign velocities in order to maintain a set target temperature of the entire system. Widely used choices are Nosé–Hoover [72,73] or Berendsen thermostats [74], which rescale velocities according to different coupling strengths with respect to a heat bath.

Implicit solvent simulations at constant temperature may be achieved in a similar way by coupling the solute to a heat bath. However, the choice of the thermostat has implications for the kinetics of conformational sampling by the solute. When an explicit solvent system is coupled to a thermostat via a straightforward velocity rescaling scheme such as Nosé–Hoover dynamics, it is in fact possible (and likely) that the kinetic energy of the solute alone fluctuates significantly as long as the kinetic energy of the solvent fluctuates in a compensatory fashion. On the molecular level, fluctuations of kinetic energy between solute and solvent are the result of atomic collisions that may occasionally transfer significant amounts of kinetic energy. Because implicit solvent simulations lack this possibility, direct coupling of only the solute with Nosé–Hoover or Berendsen thermostats would therefore lead to reduced fluctuations in kinetic energy compared to the explicit solvent system. The consequence is a reduced ability to overcome significant kinetic barriers as shown recently in a comparison of alanine dipeptide transitions with implicit and explicit solvent [2].

More realistic kinetic behavior in implicit solvent simulations can be obtained with the Langevin thermostat [18] where stochastic collisions and friction forces provide kinetic energy transfer to and from the solute in an analogous fashion to explicit solute–solvent interactions. As a result, kinetic transition rates similar to rates from explicit solvent simulations can be recovered with an appropriate choice of the friction constant [2].

Often, the primary objective of implicit solvent simulations is to explore the thermodynamics of the accessible conformational space rather than kinetics. In that case, Langevin dynamics with reduced friction coefficients may be used to accelerate the rate of kinetic barrier crossings [2,3]. The sampling of conformational basins without significant barriers (compared to the thermal energy, kT) is also accelerated by Langevin dynamics with reduced friction coefficients. Interestingly, simulations with the Nosé–Hoover thermostat actually cover conformational space even more efficiently in the absence of barriers due to the lack of friction and the lack of stochastic collisions which randomize motion as a result of atomic velocity reassignments [2]. Consequently, the thermostat is an important choice in implicit solvent simulations depending on the type of problem that is being addressed.

3.3 Crowded cellular environments

As a first approximation, crowded cellular environments may be modeled as a dielectric continuum with a reduced effective dielectric constant (Figure 6.1A), possibly in combination with an increased cost of cavity formation to account for the

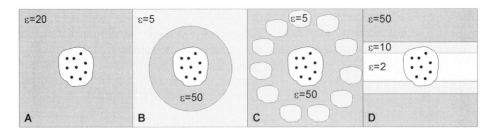

FIGURE 6.1   Proposed continuum electrostatic models of cellular environments.

steric constraints on insertion of a solute due to a high concentration of surrounding macromolecules. The effective solvent polarization experienced by a solute in such an environment is reduced because water is partially immobilized [17] when sandwiched between biomolecules and because less polarizable macromolecules in the vicinity lower the overall dielectric response due to the long-range nature of electrostatic interactions. While the average solvent polarizability of cellular environments may be highly variable and is difficult to determine experimentally, one may expect the dielectric constant to fall in a range between 30 and 50 [75–77].

Initial studies of peptide conformational sampling in reduced dielectric environments indicate that even a moderate reduction of the dielectric constant can significantly alter conformational preferences [78]. The reduced electrostatic screening in lower dielectric environments generally favors interactions between charged and polar groups vs. hydrophobic core formation in high dielectric environments. This may result in increased helical propensities at low dielectric constants in a system like poly-alanine but shift conformational equilibria and aggregation propensities in a more complex manner in amphiphilic peptides such as melittin [78].

An implicit representation of more complex heterogeneous cellular environments is also possible by assigning different dielectric constants to different spatial domains and determining an effective dielectric constant throughout space from solutions to the Poisson equation with a probe sphere according to Eq. (5). One may for example assume a system consisting of a high dielectric constant in a spherical cavity surrounded by a low dielectric environment (see Figure 6.1B). This would be reflective of a hydrated solute with a few solvation layers but surrounded by a shell of macromolecules. In this case, the resulting effective dielectric would be a radial function that only depends on the distance from the center of the high dielectric sphere. The model could be refined further by embedding spherical low dielectric cavities into a high dielectric continuum as shown in Figure 6.1C and recalculating an effective dielectric function that now varies three-dimensionally. The model could be augmented with an exclusion potential that keeps a solute of interest from entering the low dielectric cavities where it would overlap with another macromolecule. In reality, the environment is not fixed, but an independent dynamic variation of the low dielectric cavities would require a computationally very expensive recalculation of the effective dielectric profile at every simulation step. A dynamic variation of the dielectric profile becomes feasible in principle, however, if the profile can be altered through a few degrees of freedom such as stretching of the profile in radial direction to reflect a smaller or larger distance of the surrounding macromolecules from the center of the system.

3.4 Membrane bilayers

Biological membranes are a special case of heterogeneous cellular environments. According to the physicochemical characteristics of the constituent ordered phospholipids they consist of a hydrophobic interior, a layer of intermediate polarizability and a polar, zwitterionic, or charged head group region with interspersed water molecules. Such a system could be approximated with a simple two dielectric model consisting of a low dielectric slab embedded in a high dielectric environment. Such a model can be simplified further by assuming that the dielectric constant of the hydrophobic layer is equal to the internal dielectric of the solute (usually 1) which greatly simplifies the application of existing GB formalisms [79–81]. A more realistic implicit model of a membrane bilayer would consist of multiple dielectric layers with the interior hydrophobic layer exhibiting at least some polarizability with ε between 1 and 2 (see Figure 6.1D) [82]. Such a model results in a smooth effective dielectric profile which can be calculated for a specific set of layers according to Poisson theory [82] or may be approximated as a simple Gaussian function [83]. The exact choice of dielectric constants and width of the dielectric layers depends on the lipid type and is subject to parameterization with respect to reference data from experiments and explicit lipid simulations.

The non-polar component of the solvation free energy is especially important for implicit membrane models as it decreases from a significant positive contribution in aqueous solvent to near zero at the center of the phospholipid bilayer. Without a non-polar term, even hydrophobic solutes would in fact prefer the high-dielectric environment where the electrostatic solvation free energy is more favorable than in a low-dielectric medium. The functional form of the non-polar term may follow a simple switching function [79,80], a calculated free energy insertion profile for molecular oxygen [82,84], or may be parameterized as well with respect to simulation or experimental data.

Particularly useful data for parameterization of implicit membrane models are experimental transfer free energies between water and cyclohexane [85] (as a mimic of the hydrophobic tails) and explicit lipid free energy calculations for the insertion of amino acid side chain analogs [86,87]. Insertion profiles for amino acid side chain analogs with implicit and explicit models are shown in Figure 6.2. The optimized effective dielectric and non-polar profiles for use with the HDGB (heterogeneous dielectric generalized Born) implicit model [9,82] are shown in Figure 6.3.

The agreement between the implicit and explicit model and between the calculated transfer free energies and the experimental data is quite good indicating that an implicit membrane model can provide a reasonable description of membrane insertion energetics. Similarly good agreement with experimental transfer free energies has also been found for another implicit membrane model [83]. Remaining discrepancies may be attributed to deficiencies in modeling non-polar interactions

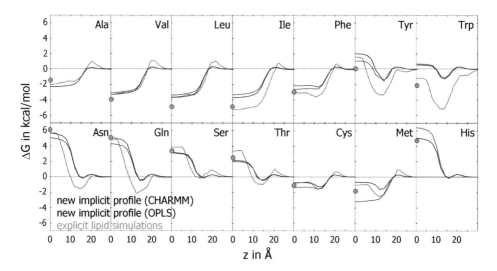

FIGURE 6.2 Amino acid side chain analog insertion profiles with explicit (red) and implicit (purple, black) membrane models. The explicit lipid profiles were calculated with the OPLS force field [87], implicit profiles were calculated with both CHARMM [92] and OPLS force fields [93]. Experimental water-cyclohexane transfer free energies [85] are indicated as red dots.

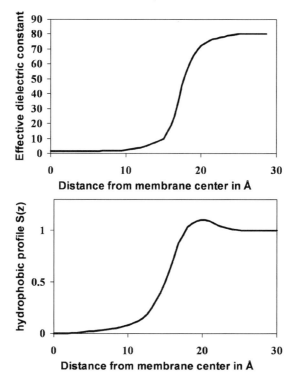

FIGURE 6.3 Optimized effective dielectric profiles and non-polar profiles to be used in implicit membrane model.

(e.g. in the case of tryptophan) and the assumption of a fixed membrane width that does not allow water defects at the membrane–water interface near polar residues as seen in the explicit lipid simulations [87].

Implicit membrane models have been used successfully in simulation of membrane proteins [9] and in folding studies of membrane-bound peptides [88–90]. Furthermore, applications of implicit membrane models as scoring functions and in MMGB/SA-type free energy calculations for membrane-bound biomolecules are conceivable [91].

4. SUMMARY AND OUTLOOK

Over the recent years implicit solvent models have undergone a transition to relatively mature methodology that is now widely employed in molecular dynamics simulations and related applications. Most popular are implicit solvent models based on a decomposition of the solvation free energy into electrostatic and nonpolar components. The electrostatic free energy is typically obtained according to a continuum electrostatics model that is described by Poisson theory or by the more approximate but much more efficient Generalized Born formalism.

A new direction is the extension of Generalized Born models to non-aqueous solvent environments with lowered dielectric constants and possibly a high degree of heterogeneity e.g. in biological membranes. Implicit solvent provides a mean-field picture that is useful to study the conformational sampling of biomolecules in such environments without having to consider specific interactions with specific constituents. Furthermore, the reduced representation offers computational advantages that are most pronounced for small solutes. At the same time, the lack of the specific interactions with the environment is the greatest disadvantage in cases where such interactions play an important role. Future developments may therefore focus on the development of effective hybrid methodology where only selected solvent molecules are represented explicitly in an otherwise implicit environment.

Another direction that needs to be addressed is the rather limiting assumption that cellular environments are largely static. Variations in the effective dielectric constant as a function of both space and time are possible in principle. Future efforts may address how to parameterize and implement fully variable models in an effective manner.

Finally, implicit models of complex cellular environments are the first step towards developing a comprehensive multi-scale modeling framework for simulating cellular processes in molecular detail and address future challenges in molecular biology.

ACKNOWLEDGMENTS

Financial support from National Science Foundation CAREER Grant 0447799 and the Alfred P. Sloan Foundation is acknowledged.

REFERENCES

1. Roux, B., Simonson, T. Implicit solvent models. Biophys. Chem. 1999, 78, 1–20.
2. Feig, M. Kinetics from implicit solvent simulations of biomolecules as a function of viscosity. J. Chem. Theory Comput. 2007, 3, 1734–48.
3. Zagrovic, B., Pande, V. Solvent viscosity dependence of the folding rate of a small protein: Distributed computing study. J. Comput. Chem. 2003, 24, 1432–6.
4. Sugita, Y., Okamoto, Y. Replica-exchange molecular dynamics method for protein folding. Chem. Phys. Lett. 1999, 314, 141–51.
5. Feig, M., Brooks III, C.L. Recent advances in the development and application of implicit solvent models in biomolecule simulations. Curr. Opin. Struct. Biol. 2004, 14, 217–24.
6. Gnanakaran, S., Nymeyer, H., Portman, J., Sanbonmatsu, K.Y., Garcia, A.E. Peptide folding simulations. Curr. Opin. Struct. Biol. 2003, 13, 168–74.
7. Im, W., Chen, J.H., Brooks, C.L. Peptide and protein folding and conformational equilibria: Theoretical treatment of electrostatics and hydrogen bonding with implicit solvent models. Adv. Protein Chem. 2006, 72, 173.
8. Onufriev, A., Bashford, D., Case, D.A. Exploring protein native states and large-scale conformational changes with a modified Generalized Born model. Proteins 2004, 55, 383–94.
9. Tanizaki, S., Feig, M. Molecular dynamics simulations of large integral membrane proteins with an implicit membrane model. J. Phys. Chem. B 2006, 110, 548–56.
10. Ruscio, J.Z., Onufriev, A. A computational study of nucleosomal DNA flexibility. Biophys. J. 2006, 91, 4121–32.
11. Ewald, P.P. Die Berechnung optischer und elektrostatischer Gitterpotentiale. Ann. Physik 1921, 64, 253–87.
12. Darden, T.A., York, D., Pedersen, L.G. Particle mesh Ewald: An Nlog(N) method for Ewald sums in large systems. J. Chem. Phys. 1993, 98, 10089–92.
13. Lee, M.S., Salsbury Jr., F.R., Brooks III, C.L. Novel Generalized Born methods. J. Chem. Phys. 2002, 116, 10606–14.
14. Krol, M. Comparison of various implicit solvent models in molecular dynamics simulations of immunoglobulin G light chain dimer. J. Comput. Chem. 2003, 24, 531–46.
15. Zimmerman, S.B., Minton, A.P. Macromolecular crowding—Biochemical, biophysical, and physiological consequences. Annu. Rev. Biophys. Biomol. Struct. 1993, 22, 27–65.
16. Friedel, M., Sheeler, D.J., Shea, J.-E. Effects of confinement and crowding on the thermodynamics and kinetics of folding of a minimalist β-barrel protein. J. Chem. Phys. 2003, 118, 8106–13.
17. Despa, F., Fernandez, A., Berry, R.S. Dielectric modulation of biological water. Phys. Rev. Lett. 2004, 93.
18. Brooks, C.L., Berkowitz, M., Adelman, S.A. Generalized Langevin theory for many-body problems in chemical-dynamics—Gas-surface collisions, vibrational-energy relaxation in solids, and recombination reactions in liquids. J. Chem. Phys. 1980, 73, 4353–64.
19. Wesson, L., Eisenberg, D. Atomic solvation parameters applied to molecular dynamics of proteins in solution. Protein Sci. 1992, 1, 227–35.
20. Lazaridis, T., Karplus, M. Effective energy function for proteins in solution. Proteins 1999, 35, 133–52.
21. Ferrara, P., Apostolakis, J., Caflish, A. Evaluation of a fast implicit solvent model for molecular dynamics simulations. Proteins 2002, 46, 24–33.
22. Sharp, K.A., Honig, B. Electrostatic interactions in macromolecules: Theory and applications. Annu. Rev. Biophys. Chem. 1990, 19, 301–32.
23. Gilson, M.K., Sharp, K.A., Honig, B.H. Calculating the electrostatic potential of molecules in solution: Method and error assessment. J. Comput. Chem. 1987, 9, 327–35.
24. Warwicker, J., Watson, H.C. Calculation of the electric potential in the active site cleft due to α-helix dipoles. J. Mol. Biol. 1982, 157, 671–9.
25. Feig, M., Onufriev, A., Lee, M.S., Im, W., Case, D.A., Brooks III, C.L. Performance comparison of Generalized Born and Poisson methods in the calculation of electrostatic solvation energies for protein structures. J. Comput. Chem. 2004, 25, 265–84.

26. Zhou, Y.C., Feig, M., Wei, G.W. Highly accurate biomolecular electrostatics in continuum dielectric environments. J. Comput. Chem. 2008, 29, 87–97.
27. Still, W.C., Tempczyk, A., Hawley, R.C., Hendrickson, T. Semianalytical treatment of solvation for molecular mechanics and dynamics. J. Am. Chem. Soc. 1990, 112, 6127–9.
28. Sitkoff, D., Sharp, K.A., Honig, B. Accurate calculation of hydration free-energies using macroscopic solvent models. J. Phys. Chem. 1994, 98, 1978–88.
29. Tan, C., Tan, Y.H., Luo, R. Implicit nonpolar solvent models. J. Phys. Chem. B 2007, 111, 12263–74.
30. Wagoner, J.A., Baker, N.A. Assessing implicit models for nonpolar mean solvation forces: The importance of dispersion and volume terms. Proc. Natl. Acad. Sci. USA 2006, 103, 8331–6.
31. Gallicchio, E., Levy, R.M. AGBNP: An analytic implicit solvent model suitable for molecular dynamics simulations and high-resolution modeling. J. Comput. Chem. 2004, 25, 479–99.
32. Chen, J., Brooks III, C.L. Implicit modeling of nonpolar solvation for simulating protein folding and conformational transitions. Phys. Chem. Chem. Phys. 2008, 10, 471–81.
33. Stern, H.A., Feller, S.E. Calculation of the dielectric permittivity profile for a nonuniform system: Application to a lipid bilayer simulation. J. Chem. Phys. 2003, 118, 3401–12.
34. Sigalov, G., Scheffel, P., Onufriev, A. Incorporating variable dielectric environments into the Generalized Born model. J. Chem. Phys. 2005, 122.
35. Feig, M., Im, W., Brooks III, C.L. Implicit solvation based on Generalized Born theory in different dielectric environments. J. Chem. Phys. 2004, 120, 903–11.
36. Kirkwood, J.G. Theory of solutions of molecules containing widely separated charges with special application to zwitterions. J. Chem. Phys. 1934, 2, 351–61.
37. Bashford, D., Case, D.A. Generalized Born models of macromolecular solvation effects. Ann. Rev. Phys. Chem. 2000, 51, 129–52.
38. Lee, M.S., Feig, M., Salsbury, F.R. Jr., Brooks III, C.L. New analytical approximation to the standard molecular volume definition and its application to Generalized Born calculations. J. Comput. Chem. 2003, 24, 1348–56.
39. Brown, F.L.H., Brannigan, G. A consistent model for thermal fluctuations and protein-induced deformations in lipid bilayers. Biophys. J. 2006, 90, 1501–20.
40. Feig, M., Brooks III, C.L. Evaluating CASP4 predictions with physical energy functions. Proteins 2002, 49, 232–45.
41. Kollman, P.A., Massova, I., Reyes, C., Kuhn, B., Huo, S., Chong, L., Lee, M., Lee, T., Duan, Y., Wang, W., Donini, O., Cieplak, P., Srinivasan, J., Case, A.D., Cheatham III, T.E. Calculating structures and free energies of complex molecules: Combining molecular mechanics and continuum models. Accounts Chem. Res. 2000, 33, 889–97.
42. Li, X.F., Hassan, S.A., Mehler, E.L. Long dynamics simulations of proteins using atomistic force fields and a continuum representation of solvent effect: Calculation of structural and dynamic properties. Proteins 2005, 60, 464–84.
43. Huang, A., Stultz, C.M. Conformational sampling with implicit solvent models: Application to the PHF6 peptide in tau protein. Biophys. J. 2007, 92, 34–45.
44. Fan, H., Mark, A.E., Zhu, J., Honig, B. Comparative study of Generalized Born models: Protein dynamics. Proc. Natl. Acad. Sci. USA 2005, 102, 6760–4.
45. Watanabe, Y.S., Kim, J.G., Fukunishi, Y., Nakamura, H. Free energy landscapes of small peptides in an implicit solvent model determined by force-biased multicanonical molecular dynamics simulation. Chem. Phys. Lett. 2004, 400, 258–63.
46. Baumketner, A., Shea, J.E. The thermodynamics of folding of a beta hairpin peptide probed through replica exchange molecular dynamics simulations. Theor. Chem. Acc. 2006, 116, 262–73.
47. Bursulaya, B.D., Brooks III, C.L. Comparative study of the folding free-energy landscape of a three-stranded β-sheet protein with explicit and implicit solvent models. J. Phys. Chem. B 2000, 104, 12378–83.
48. Yoda, T., Sugita, Y., Okamoto, Y. Cooperative folding mechanism of a beta-hairpin peptide studied by a multicanonical replica-exchange molecular dynamics simulation. Proteins 2007, 66, 846–59.
49. Chen, J.H., Im, W.P., Brooks, C.L. Balancing solvation and intramolecular interactions: Toward a consistent Generalized Born force field. J. Am. Chem. Soc. 2006, 128, 3728–36.
50. Zagrovic, B., Sorin, E.J., Pande, V. β-Hairpin folding simulations in atomistic detail using an implicit solvent model. J. Mol. Biol. 2001, 313, 151–69.

51. Chen, C.J., Xiao, Y. Molecular dynamics simulations of folding processes of a beta-hairpin in an implicit solvent. Phys. Biol. 2006, 3, 161–71.
52. Nymeyer, H., Garcia, A.E. Simulation of the folding equilibrium of α-helical peptides: A comparison of the Generalized Born approximation with explicit solvent. Proc. Natl. Acad. Sci. USA 2003, 100, 13934–9.
53. Karanicolas, J., Brooks III, C.L. Integrating folding kinetics and protein function: Biphasic kinetics and dual binding specificity in a WW domain. Proc. Natl. Acad. Sci. USA 2004, 101, 3432–7.
54. Jagielska, A., Scheraga, H.A. Influence of temperature, friction, and random forces on folding of the B-domain of staphylococcal protein A: All-atom molecular dynamics in implicit solvent. J. Comput. Chem. 2007, 28, 1068–82.
55. Yang, M., Yordanov, B., Levy, Y., Brüschweiler, R., Huo, S. The sequence-dependent unfolding pathway plays a critical role in the amyloidogenicity of transthyretin. Biochemistry 2006, 45, 11992–2002.
56. Formaneck, M.S., Cui, Q. The use of a Generalized Born model for the analysis of protein conformational transitions: A comparative study with explicit solvent simulations for chemotaxis Y protein (CheY). J. Comput. Chem. 2006, 27, 1923–43.
57. Onufriev, A., Case, D.A., Bashford, D. Structural details, pathways, and energetics of unfolding apomyoglobin. J. Mol. Biol. 2003, 325, 555–67.
58. Murcia, M., Faraldo-Gomez, J.D., Maxfield, F.R., Roux, B. Modeling the structure of the StART domains of MLN64 and StAR proteins in complex with cholesterol. J. Lipid Res. 2006, 47, 2614–30.
59. Chocholousova, J., Feig, M. Implicit solvent simulations of DNA and DNA–protein complexes: Agreement with explicit solvent vs. experiment. J. Phys. Chem. B 2006, 110, 17240–51.
60. Tsui, V., Case, D.A. Molecular dynamics simulations of nucleic acids with a Generalized Born solvation model. J. Am. Chem. Soc. 2000, 122, 2489–98.
61. Feig, M., Chocholousova, J., Tanizaki, S. Extending the horizon: Towards the efficient modeling of large biomolecular complexes in atomic detail. Theor. Chem. Acc. 2006, 116, 194–205.
62. Gong, X.Q., Zhang, C.F., Feig, M., Burton, Z.F. Dynamic error correction and regulation of downstream bubble opening by human RNA polymerase II. Mol. Cell 2005, 18, 461–70.
63. Hajjar, E., Perahia, D., Débat, H., Nespoulous, C., Robert, C.H. Odorant binding and conformational dynamics in the odorant-binding protein. J. Biol. Chem. 2006, 281, 29929–37.
64. Mongan, J., Case, D.A. Biomolecular simulations at constant pH. Curr. Opin. Struct. Biol. 2005, 15, 157–63.
65. Lee, M.S., Salsbury, F.R.J., Brooks III, C.L. Constant pH molecular dynamics using continuous titration coordinates. Proteins 2004, 56, 738–52.
66. Khandogin, J., Chen, J.H., Brooks III, C.L. Exploring atomistic details of pH-dependent peptide folding. Proc. Natl. Acad. Sci. USA 2006, 103, 18546–50.
67. Chen, Y.Z., Chen, X., Deng, Y.F. Simulating botulinum neurotoxin with constant pH molecular dynamics in Generalized Born implicit solvent. Comput. Phys. Commun. 2007, 177, 210–3.
68. Chen, J.H., Won, H.S., Im, W., Dyson, H.J., Brooks III, C.L. Generation of native-like protein structures from limited NMR data, modern force fields and advanced conformational sampling. J. Biomol. NMR 2005, 31, 59–64.
69. Chen, J., Brooks III, C.L. Can molecular dynamics simulations provide high-resolution refinement of protein structure? Proteins 2007, 67, 922–30.
70. Roe, D.R., Okur, A., Wickstrom, L., Hornak, V., Simmerling, C. Secondary structure bias in Generalized Born solvent models: Comparison of conformational ensembles and free energy of solvent polarization from explicit and implicit solvation. J. Phys. Chem. B 2007, 111, 1846–57.
71. Geney, R., Layten, M., Gomperts, R., Hornak, V., Simmerling, C. Investigation of salt bridge stability in a Generalized Born solvent model. J. Chem. Theory Comput. 2006, 2, 115–27.
72. Hoover, W.G., Aoki, K., Hoover, C.G., De Groot, S.V. Time-reversible deterministic thermostats. Physica D 2004, 187, 253–67.
73. Nose, S. A molecular dynamics method for simulations in the canonical ensemble. Mol. Phys. 1984, 52, 255–68.
74. Gunsteren, W.F.V., Berendsen, H.J.C. Computer simulation of molecular dynamics: Methodology, applications, and perspectives in chemistry. Ang. Chemie 1990, 29, 992–91023.
75. Wyman, J. Studies on the dielectric constant of protein solutions. I. Zein. J. Biol. Chem. 1930, 90, 443–76.

76. Asami, K., Hanai, T., Koizumi, N. Dielectric properties of yeast cells. J. Membrane Biol. 1976, 28, 169–80.
77. Rufus, E., Alex, Z.C., Mathew, L. Dielectric relaxation studies of biological tissues in the microwave frequency range, APMC 2005 Proceedings, 2005, 16062–5.
78. Tanizaki, S., Clifford, J.W., Connelly, B.D., Feig, M. Conformational sampling of peptides in cellular environments. Biophys. J. 2008, 94, 747–59.
79. Im, W., Feig, M., Brooks III, C.L. An implicit membrane Generalized Born theory for the study of structure, stability, and interactions of membrane proteins. Biophys. J. 2003, 85, 2900–18.
80. Spassov, V.Z., Yan, L., Szalma, S. Introducing an implicit membrane in Generalized Born/solvent accessibility continuum solvent models. J. Phys. Chem. B 2002, 106, 8726–38.
81. Kessel, A., Cafiso, D.S., Ben-Tal, N. Continuum solvent model calculations of alamethicin-membrane interactions: Thermodynamic aspects. Biophys. J. 2000, 78, 571–83.
82. Tanizaki, S., Feig, M. A Generalized Born formalism for heterogeneous dielectric environments: Application to the implicit modeling of biological membranes. J. Chem. Phys. 2005, 122, 12470–6.
83. Ulmschneider, M.B., Ulmschneider, J.P., Sansom, M.S.P., Di Nola, A. A Generalized Born implicit-membrane representation compared to experimental insertion free energies. Biophys. J. 2007, 92, 2338–49.
84. Marrink, S.J., Berendsen, H.J.C. Permeation process of small molecules across lipid membranes studied by molecular dynamics simulations. J. Phys. Chem. 1996, 100, 16729–38.
85. Radzicka, A., Wolfenden, R. Comparing the polarities of the amino-acids—Side-chain distribution coefficients between the vapor-phase, cyclohexane, 1-octanol, and neutral aqueous-solution. Biochemistry 1988, 27, 1664–70.
86. Bemporad, D., Essex, J.W., Luttmann, C. Permeation of small molecules through a lipid bilayer: A computer simulation study. J. Phys. Chem. B 2004, 108, 4875–84.
87. MacCallum, J.L., Bennett, W.F.D., Tieleman, D.P. Partitioning of amino acid side chains into lipid bilayers: Results from computer simulations and comparison to experiment. J. General Phys. 2007, 129, 371–7.
88. Ulmschneider, J.P., Ulmschneider, M.B., Di Nola, A. Monte Carlo folding of trans-membrane helical peptides in an implicit Generalized Born membrane. Proteins 2007, 69, 297–308.
89. Im, W., Brooks, C.L. Interfacial folding and membrane insertion of designed peptides studied by molecular dynamics simulations. Proc. Natl. Acad. Sci. USA 2005, 102, 6771–6.
90. Im, W., Brooks, C.L. De novo folding of membrane proteins: An exploration of the structure and NMR properties of the fd coat protein. J. Mol. Biol. 2004, 337, 513–9.
91. Lomize, A.L., Pogozheva, I.D., Lomize, M.A., Mosberg, H.I. Positioning of proteins in membranes: A computational approach. Protein Sci. 2006, 15, 1318–33.
92. MacKerell Jr., A.D., Bashford, D., Bellott, M., Dunbrack, J.D., Evanseck, M.J., Field, M.J., Fischer, S., Gao, J., Guo, H., Ha, S., Joseph-McCarthy, D., Kuchnir, L., Kuczera, K., Lau, F.T.K., Mattos, C., Michnick, S., Ngo, T., Nguyen, D.T., Prodhom, B., Reiher, W.E., Roux, B., Schlenkrich, M., Smith, J.C., Stote, R., Straub, J., Watanabe, M., Wiorkiewicz-Kuczera, J., Yin, D., Karplus, M. All-atom empirical potential for molecular modeling and dynamics studies of proteins. J. Phys. Chem. B 1998, 102, 3586–616.
93. Jorgensen, W.L., Maxwell, D.S., Tirado-Rives, J. Development and testing of the OPLS all-atom force field on conformational energetics and properties of organic liquids. J. Am. Chem. Soc. 1996, 118, 11225–36.

Section 3
Simulation Methodologies

Section Editor: Carlos Simmerling

Center for Structural Biology
Stony Brook University
Stony Brook, NY 11794
USA

CHAPTER 7

Implicit Solvent Models in Molecular Dynamics Simulations: A Brief Overview

Alexey Onufriev*

Contents

1. INTRODUCTION

The effects of solvent environment must be taken into account for realistic modeling of bio-molecules. Traditionally, this has been accomplished by placing a sufficiently large number of individual water molecules around the solute, and simulating their motion on an equal footing with the molecule of interest. While arguably the most realistic of the current theoretical approaches, this *explicit solvent* methodology suffers from considerable computational costs, which often become prohibitive, especially for large systems or long time-scales, such as those involved in the folding of proteins. Other problems with the approach include the difficulty, and often inability to calculate relative free energies of molecular conformations due to the need to account for very large number of solvent degrees of freedom.

An alternative that is becoming more and more popular—the *implicit solvent model* [1–7]—is based on replacing real water environment consisting of discrete

* Department of Computer Science and Physics, 2050 Torgersen Hall, Virginia Tech, Blacksburg, VA 24061, USA

Annual Reports in Computational Chemistry, Vol. 4
ISSN 1574-1400, DOI: 10.1016/S1574-1400(08)00007-8

molecules by an infinite continuum medium with the dielectric and "hydropho-bic" properties of water. Presented below is a very brief overview of the cur-rent state of the methodology, with specific focus on the hierarchy of underlying approximations and the use of corresponding computational models in molec-ular dynamics (MD) simulations. One specific example—the Generalized Born model—is discussed in relatively greater detail, reflecting the author's own ex-perience with the development of this model.

2. IMPLICIT SOLVENT FRAMEWORK

Implicit solvent models have several advantages over explicit water representa-tions, especially in molecular dynamics simulations. These include the following.

Lower computational costs for many molecular systems, and better scaling on parallel machines. The effective cost reduction may be particularly significant if one takes into account the improved sampling: in contrast to explicit solvent mod-els, solvent viscosity that slows down conformational transitions can be turned off completely within implicit representations.

Effective ways to estimate free energies; since solvent degrees of freedom are taken into account implicitly, estimating free energies of solvated structures is much more straightforward than with explicit water models.

Since implicit solvent models correspond to instantaneous dielectric response from solvent, there is no need for the lengthy equilibration of water that is typically necessary in explicit water simulations. This feature of implicit solvent models becomes key when charge state of the system is changed many times during the course of a simulation, as, for example, in constant pH simulations.

Finally, the implicit solvent approach has a clear advantage over explicit sol-vent in computing and making physical sense of energy landscapes of molecular structures. Here, implicit averaging over solvent degrees of freedom eliminates the "noise"—an astronomical number of local minima arising from small variations in solvent structure.

2.1 The hierarchy of underlying approximations

When contemplating a use of practical techniques based on the implicit solvent framework, one should be keenly aware of the fact that all of the attractive fea-tures of the methodology listed above come at a price of making a number of approximations whose effects are often hard, if not impossible, to estimate. Note that the *discrete* → *continuum* step is not the only deviation from reality: as one descends the "tree of approximations" that the methodology is based upon, Fig-ure 7.1, down to models used in practice today, more and more approximations are made. Also note that some familiar descriptors of molecular interaction, such as solute–solvent hydrogen bonds, are no longer explicitly present in the model—instead, they come in implicitly, albeit at a mean-field level, and contribute to the overall solvation energy.

In many molecular modeling applications, and especially in molecular dy-namics, the key quantity that needs to be computed is the total energy of the

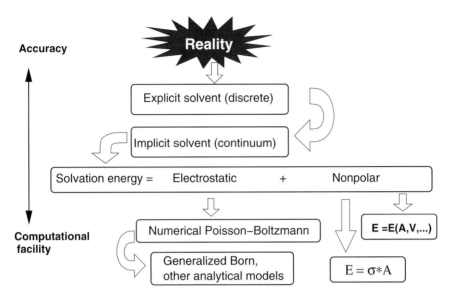

FIGURE 7.1 An "approximations tree" of the implicit solvent framework.

molecule in the presence of solvent. This energy is a function of molecular configuration, its gradients with respect to atomic positions determine the forces on
the atoms. The total energy of a solvated molecule can be conveniently written as
$E_{tot} = E_{vac} + \Delta G_{solv}$, where E_{vac} represents molecule's potential energy in vacuum
(gas-phase), and ΔG_{solv} is defined as the free energy of transferring the molecule
from vacuum into solvent, i.e. solvation free energy.[1] In practice, once the choice
of the gas-phase potential function, or force-field, E_{vac} is made, its computation
is relatively straightforward [8]. The difficulty comes from the need to estimate
the effects of solvent, encapsulated by the ΔG_{solv} term in the above equation. At
present, the implicit solvent framework makes the following simplifying approximation to estimate ΔG_{solv}:

$$\Delta G_{solv} = \Delta G_{el} + \Delta G_{nonpolar}, \tag{1}$$

where $\Delta G_{nonpolar}$ is the free energy of solvating a molecule from which all charges
have been removed (i.e. partial charges of every atom are set to zero), and ΔG_{el}
is the free energy of first removing all charges in the vacuum, and then adding
them back in the presence of a continuum solvent environment. To proceed, one
needs practical methods of computing both ΔG_{el} and $\Delta G_{nonpolar}$. The "hydrophobic" part $\Delta G_{nonpolar}$ represents the combined effect of two types of interaction:
the favorable van der Waals attraction between the solute and solvent molecules,
and the unfavorable cost of breaking the structure of the solvent (water) around
the solute. The common approximation widely in use today [9] assumes both of
these contributions to be proportional to the total solvent accessible surface area

---

[1] Technically, the above decomposition is already an approximation made by most classical (non-polarizable) force-
fields, as it assumes this specific separability of the Hamiltonian.

(A) of the molecule, thus taking $\Delta G_{nonpolar} \approx \sigma \times A$, with the proportionality constant derived from experimental solvation energies of small non-polar molecules. Substantial uncertainty exists in what appropriate value of the surface tension σ should be used in simulations, which perhaps reflects the limitations of this approximation itself. Strong arguments for the use of less drastic approximations for $\Delta G_{nonpolar}$, e.g. those that treat solute-solvent van der Waals interactions ("volume term") separately from the surface area term, have also been made [10,11]. At the same time, some researchers choose to neglect the surface area term altogether in MD simulations, especially if no large conformational changes are expected, for example in simulations of proteins in their native states. Regardless of the specific form, computing the hydrophobic term has not so far been the computational bottleneck of a typical MD simulation. Currently, the most time-consuming part is the computation of the electrostatic contribution to the total solvation free energy, ΔG_{el}. The underlying long-range interactions are critical to function and stability of many classes of biological and chemical structures, and so it is not surprising that considerable effort was put into making these computations accurate and fast.

2.2 The Poisson–Boltzmann model

If one accepts the continuum, linear response dielectric approximation for the solvent, then the Poisson equation of classical electrostatics provides an exact formalism for computing the electrostatic potential $\phi(\mathbf{r})$ produced by a molecular charge distribution $\rho(\mathbf{r})$. The screening effects of salt can be added at this level via an approximate mean-field treatment, resulting in the so-called *Poisson–Boltzmann* (PB) equation [13]. In general, this is a second order non-linear partial differential equation, but its simpler linearized form is often used in biomolecular applications:

$$\nabla\big[\epsilon(\mathbf{r})\nabla\phi(\mathbf{r})\big] = -4\pi\rho(\mathbf{r}) + \kappa^2\epsilon(\mathbf{r})\phi(\mathbf{r}), \qquad (2)$$

where $\epsilon(\mathbf{r})$ is the position-dependent dielectric constant, and the screening effects of salt enter via the Debye–Hückel parameter $\kappa \sim \sqrt{[salt]}$. Once the $\phi(\mathbf{r})$ for a given molecular configuration is obtained via numerical solution of the PB equation, the electrostatic part of the solvation energy, ΔG_{el}, can be computed. Details of numerical procedures for solving the PB equation along with a discussion of some of the related technical issues, can be found in recent literature on the subject, e.g. [12] and references therein. While the numerical PB formalism has been successfully applied to "static" structures for the past 20 years, it was not until quite recently that its use in MD simulations has been reported. In part, this delay was due to the relatively high costs associated with solving the PB equation at every MD step. Technical difficulties associated with computing forces due to dielectric boundary had to be overcome as well. So far, very few, mainly "proof-of-concept" PB-based MD simulations have been reported [7,14–16]. Still, the approach holds tremendous potential for MD simulations [13]. This is because the PB model has a rigorous physical basis and requires fewer fundamental approximations to physical reality than most other implicit solvent approaches currently in use. The model also serves as a natural reference point on the "approximations tree," Figure 7.1. Comparisons of results from PB-based simulations with those based on the more fundamental explicit solvent model helps reveal basic limitations of the implicit solvent approach itself, while comparisons with even more approximate methods, such as the widely used generalized Born model discussed below, help assess the accuracy of the latter [17].

2.3 The Generalized Born model

The need for computationally facile models for dynamical applications requires further trade-offs between accuracy and speed. Descending from the PB model down the approximations tree, Figure 7.1, one arrives at the generalized Born (GB) model that has been developed as a computationally efficient approximation to numerical solutions of the PB equation. The analytical GB method is an approximate, relative to the PB model, way to calculate the electrostatic part of the solvation free energy, ΔG_{el}, see [18] for a review. The methodology has become particularly popular in MD applications [10,19–23], due to its relative simplicity and computational efficiency, compared to the more standard numerical solution of the Poisson–Boltzmann equation.

2.3.1 The underlying approximations of the GB model
GB models evaluate electrostatic part of solvation free energy as a sum of pairwise interaction terms between atomic charges. For a typical case of aqueous solvation of molecules with interior dielectric of 1, these interactions are approximated by an analytical function introduced by Still et al. [24]:

$$\Delta G_{el} \approx -\frac{1}{2}\left(1 - \frac{1}{\epsilon_w}\right) \sum_{i,j} \frac{q_i q_j}{\sqrt{r_{ij}^2 + R_i R_j \exp(-\frac{r_{ij}^2}{4R_i R_j})}} \qquad (3)$$

where r_{ij} is the distance between atoms i and j, q_i and q_j are partial charges and $\epsilon_w \gg 1$ is the dielectric constant of the solvent. Screening effects of monovalent salt can be introduced at the Debye–Hückel level by a simple, computationally inexpensive empirical correction to the above equation [25].

The key parameters in the GB function are the effective Born radii of the interacting atoms, R_i and R_j, which represent each atom's degree of burial within the solute. More specifically, the effective radius of an atom is defined as the radius of a corresponding spherical ion having the same ΔG_{el} as would the same molecule with partial charges set to zero for all atoms except the atom of interest. Assuming that effective Born radii can be computed efficiently for every atom in the molecule, computational advantages of Eq. (3) relative to numerical PB treatment become apparent: the GB formula is simple, its analytical derivatives with respect to atomic positions immediately provide the forces. In practice, the effective radius for each atom is generally calculated by first approximating the electrostatic energy density due to the atom of interest by some reasonably simple expression and then integrating over the appropriate volume [26–32] or surface [33].

The Coulomb field approximation—CFA—is historically the first approximation of that nature. Although it makes what appears to be a fairly drastic assumption that the electric field generated by the atomic point charge is unaffected by the non-homogeneous dielectric environment created by the solute, practical routines developed on the basis of CFA are still widely used. Fortuitous cancellation of errors [34] and computational efficiency of the approximation have contributed to its success. Empirical corrections to the CFA based on multiple integrals over solute have lead to spectacular improvements in accuracy of the GB model relative to PB treatment [30,31]. Several GB models based on these approximations have been implemented in CHARMM. Recently, it was shown that the same, and possibly even better level of accuracy can be achieved with a single integral [35]. It remains to be seen whether potential advantages of this approximation [36]—termed R6—will translate into practical gains once implemented in MD codes.

Computationally effective integration over physically realistic [37], but geometrically complex molecular volume (or surface) presents a set of its own challenges: routines that match molecular volume closely, such as GBMV models [30,31] in CHARMM, typically come at a price of noticeably higher computational costs [38]. Alternative approaches include the use of physically less justified, but computationally more suitable VDW volume, combined with pairwise descreening approach [27], and empirical corrections that bring in some elements of molecular volume. Such compromise solutions [32,39], e.g. GB-OBC and GBn models in AMBER, are significantly faster, but at a cost of making additional approximations to reality.

Currently, a large variety of flavors of the GB model are available in many molecular simulation packages. The vast majority, if not all, of these models share the same foundation—Eq. (3)—but may differ substantially in the way the effective radii are computed. The algorithmic simplicity and reasonable accuracy of the GB approximation, combined with its availability in popular modeling packages, have made it the current "workhorse" in many practical applications of the implicit solvent methodology.

2.3.2 GB-based MD simulations. Examples

Protein folding. Exploring large conformational transitions is one of several areas where the advantages of implicit solvent framework, and the GB model in particular, become apparent. Several all-atom MD simulations of *ab initio* folding of small proteins have been reported. Examples include 20-residue "trpcage" protein [40], 36-residue villin headpiece [41], and a 46-residue helix bundle [42]. In these simulations the folded state was predicted to within 2 Å from experiment (C_α rmsd), and in some cases [40] within 1 Å. Energy landscapes computed within the implicit solvent framework were used to gain insights into the folding mechanisms [41,43]. Note that experimental folding times for even the fastest folding proteins is of the order of microseconds, whereas in some of the above simulations [40] the native state was reached on 10 ns time-scale. The comparison gives a very rough idea of the magnitude of conformational search speed-ups that one can expect in these types of simulations through the use of the GB approach.

Large-scale motions in proteins. The conformational search speed-up allows one to study large-scale motions in proteins and protein complexes. The use of the methodology to understand large conformational changes of the active site flaps in HIV protease [44] is a representative example: it is unlikely that a comparable explicit solvent study would currently be computationally feasible.

Membrane environment. Membranes are large structures, translocation of molecular structures through membranes may involve significant conformational changes, and so these systems are natural candidates for implicit solvent modeling. One of the challenges here is accurate and computationally facile representation of the complex dielectric environment that, in the case of membranes, includes solvent, solute, and the membrane, all with different dielectric properties. Corrections to the GB model have been introduced [45–47] to account for the effects of variable dielectric environment. Other implicit membrane models, not based on the GB, have also been proposed [48].

The DNA. Compared to proteins, implicit solvent MD simulations of nucleic acids are relatively new, and not as many. A number of methodological issues still await resolution, in particular that of appropriate treatment of multi-valent ions that are often critical for nucleic acid function. So far, the GB methodology has been employed to model free DNA in solution [49,50], binding between proteins and nucleic acids [51–53], as well as for energetic analysis of conformational changes such as the $A \rightarrow B$ transition [21]. The potential of the methodology for modeling large scale dynamics of the DNA has been demonstrated in a recent all-atom study of the nucleosome and its 147-bp DNA free in solution [54].

Constant pH simulations. The charge states of all ionizable groups remain fixed throughout the course of a typical MD simulation, regardless of the conformational changes that the structure may undergo. In reality, changes in protonation state and conformational changes are strongly coupled; this coupling may lead to non-trivial effects. To model these effects, several models have been developed. One of them employs a continuous protonation state model [55], in which equations of motion are used to time-evolve the protonation coordinate; convergence to physical protonation state of 1 or 0 is enforced by an adjustable potential barrier. An alternative approach [56] operates directly in the physical protonation space: protonation states are accepted or rejected on the fly, according to a Metropolis criterion, during the course of the MD simulation. It is the instantaneous dielectric response of the implicit solvent model that makes these on-the-fly estimates of relative energies possible.

2.3.3 Limitations of the GB model

The generalized Born model is separated from reality by several layers of approximations, Figure 7.1, each of them adding its own limitations to the method. The fundamental "discrete $\rightarrow$ continuum" approximation obviously eliminates a number of solvent effects that depend on the finite size of water molecule, such as de-wetting. Likewise, the implicit solvent model cannot describe effects of tightly bound water molecules, which may be a serious limitation when those are important for function or stability of the structure of interest. One also wonders how well the approximation works inside deep binding pockets, where solvent can hardly be considered as having properties of the bulk. Also, the additivity of ΔG_{el} and $\Delta G_{nonpolar}$ in the decomposition of total solvation free energy holds only approximately: if it were exact, absolute values of solvation energies of ions of the same size and opposite charge (and the same magnitude) would be identical, which is not the case in reality [57]. The Poisson–Boltzmann approximation inherits generic limitations of mean-field theories and linear response approximations. In particular, the neglect of correlation between counterions, especially multi-valent ones such as Mg^{2+}, may be a serious problem in the modeling of nucleic acids.

While all of these limitations are well known, their combined effect is hard, if not impossible to quantify in realistic biomolecular simulations. Understanding the effects of a single approximation, such as the $PB \rightarrow GB$ step, may be somewhat easier. Note that the GB and PB models share the same physical basis, and so one can, in principle, "derive" the GB from the PB. For example, it was recently shown that, without the heuristic exponential term, the key formula of the GB model, Eq. (3), is the limiting case ($\epsilon_w \rightarrow \infty$) of the exact solution of the Poisson equation for an arbitrary charge distribution inside an ideal sphere [36,58]. It is also possible to differentiate the effects of the $PB \rightarrow GB$ approximation from the more fundamental limitations of the PB model itself. For example, it was shown [59] that the folding landscape of β-hairpin derived from GB-based simulations is substantially different from that predicted by an explicit solvent model, which is generally more consistent with experiment. A subsequent study [60] revealed that a significant part of this discrepancy was already present at the PB level. Direct comparisons of ΔG_{el} between GB, PB, and explicit solvent are especially valuable in the context of understanding the separate effects of the approximations made. For example, it was found that even the use of "perfect" [34] effective radii in the GB Eq. (3) did not match the accuracy of the PB in predicting relative energies of poly-alanine conformational states [61]. The error of the PB itself, relative to explicit solvent treatment, was found to be smaller, but not negligible. Overall, ensembles of poly-alanine conformations generated in this study with the GB-based MD showed an overabundance of α-helical secondary structure relative to the explicit solvent results.

Raw computational speed has been considered one of the key advantages of the GB model. However, note that the cost of a calculation based directly on Eq. (3) is generally $O(N^2)$ for a system of N atoms, while the scaling is more favorable, $N \log(N)$, for Ewald-based methods used in explicit solvent simulations. For large systems, e.g. the nucleosome (25,000 atoms), the number of nanoseconds of MD per CPU hour may actually be less in a GB-based simulation (without additional approximations such as cut-offs) than in a comparable explicit solvent run [54], although the conformational search is still much faster in the implicit solvent.

2.4 Other models based on implicit solvation

While at present the GB models are arguably the most often used practical approaches in MD simulations based on implicit solvation, they are by no means the only ones. Some representative alternatives to the GB (and the PB) are listed below, in an order that roughly corresponds to their place on the "approximations tree" of Figure 7.1. In an ideal world, models that employ fewer approximations to reality may be expected to be more accurate, but this expectation may not apply to practical implementations of the methods.

Historically, MD simulations often relied on the so-called distance-dependent dielectric model [62] to account for solvation effects. In this approach, electrostatic effects are modeled by Coulomb's law with the dielectric being some fixed function of the charge–charge distance, e.g. $\epsilon(r) = r$ in the most basic form of the model. Even though the model is generally expected to be less accurate than the GB [17], its utmost simplicity and computational efficiency keep it in active use today [63,64].

Several methods for computing ΔG_{el} can be placed at roughly the same level of approximation as the GB. Examples include the generalized reaction field method [65] and the ALPB [66] model. The latter has a simple functional form similar to the GB, but contains an extra physical parameter (effective electrostatic size of the solute) and a more realistic dependence on dielectric constants. Another example of models in this group are approaches for estimation of ΔG_{el} based on image-charge solutions [67,68]. Yet another approach, AGBNP [10], combines the basic GB framework with a model for $\Delta G_{nonpolar}$ that goes beyond the surface area approximation.

At the "PB level," a model based on a very different paradigm has recently been tested in a "proof-of-concept" simulation: Maxwell's equations for the electric and magnetic field, coupled with the usual Newton's equations of motion for the charges were used to determine time-evolution of the system [69].

Going beyond the mean-field level, several "hybrid" approaches are now being explored in MD simulations. Examples include a recent model [70] in which the immediate hydration of the solute is modeled explicitly by a layer of water molecules, and the GB model is used to treat the bulk continuum solvent outside the explicit simulation volume. A similar idea was recently found very effective in the context of replica-exchange simulations [71]. An explicit ion/implicit water (PB) solvation model for molecular dynamics of nucleic acids has recently been tested [72].

Some approaches approximate the total solvation energy ΔG_{solv} without explicitly assuming additivity of the ΔG_{el} and $\Delta G_{nonpolar}$ components. One example is a Gaussian solvent-exclusion model [73] based on an empirical decomposition of ΔG_{solv} into contributions from different chemical groups. Models based on "first-principles" free energy functionals have also been proposed [74].

3. CONCLUSIONS AND OUTLOOK

An accurate description of the solvent environment is essential for realistic biomolecular modeling, but often becomes prohibitively expensive computationally if water is treated explicitly. Implicit solvent framework is an attractive alternative that offers several significant advantages over the explicit water representation, including lower computational costs, faster conformational search, and very effective ways to estimate relative free energies of conformational ensembles. However, these advantages come at a price of making several fundamental, hierarchical approximations to reality. Additional accuracy/speed trade-offs are often made in the development of computationally facile models for practical MD simulations.

Prominent among these models is the generalized Born (GB) model: although separated from reality by several layers of approximation, it apparently captures enough of the key physics of aqueous solvation to be practically useful. Compared to other models based on implicit solvation, this algorithmically simple model is arguably the one that is currently used most often in MD simulations. Many successful applications of the model to challenging problems, such as the protein folding, or the exploration of large-scale motions in proteins or DNA, have been reported. For some types of simulations, e.g. constant pH molecular dynamics, models based on implicit solvation such as the GB appear to be the only ones currently available in practice.

At the same time, examples where the GB model breaks down are also well known. Part of the overall error in these cases is attributable to the $PB \rightarrow GB$ approximation, while the remainder comes from the more fundamental limitations of the general implicit solvent framework itself. These examples are extremely important for defining the current boundaries of applicability of the GB model; they also suggest directions for future improvements.

A number of alternatives to the GB, both below and above it on the "approximations tree" have been tested in molecular dynamics simulations. Approaches that make fewer fundamental approximations to reality, such as those based directly on the Poisson–Boltzmann treatment of solvation or ones that even go beyond the mean-field level, are particularly attractive from the accuracy point of view. More testing is needed to better characterize the overall performance of these models in practical MD simulations.

In summary, the use of implicit solvation models in molecular simulations offers considerable rewards, both at conceptual and practical levels. However, compared to the more established explicit solvent approach, less is known about the domain of applicability of these models, and so extra care must be taken when using them in practice. Drawing on the analogy with the development of the empirical explicit solvent force-fields over the past 30 years, it is likely that improvements in the implicit solvent framework accompanied by accumulation of practical experience will eventually make the framework a standard approach within its reasonably well-defined domain.

ACKNOWLEDGMENTS

The author thanks Grigori Sigalov and Ramu Anandakrishnan for useful comments.

REFERENCES

1. Cramer, C.J., Truhlar, D.G. Implicit solvation models: Equilibria, structure, spectra, and dynamics. Chem. Rev. 1999, 99, 2161–200.
2. Honig, B., Nicholls, A. Classical electrostatics in biology and chemistry. Science 1995, 268, 1144–9.

3. Beroza, P., Case, D.A. Calculation of proton binding thermodynamics in proteins. Methods Enzymol. 1998, 295, 170–89.
4. Madura, J.D., Davis, M.E., Gilson, M.K., Wade, R.C., Luty, B.A., McCammon, J.A. Biological applications of electrostatic calculations and Brownian dynamics. Rev. Comput. Chem. 1994, 5, 229–67.
5. Gilson, M.K. Theory of electrostatic interactions in macromolecules. Curr. Opin. Struct. Biol. 1995, 5, 216–23.
6. Scarsi, M., Apostolakis, J., Caflisch, A. Continuum electrostatic energies of macromolecules in aqueous solutions. J. Phys. Chem. A 1997, 101, 8098–106.
7. Luo, R., David, L., Gilson, M.K. Accelerated Poisson–Boltzmann calculations for static and dynamic systems. J. Comput. Chem. 2002, 23, 1244–53.
8. Schlick, T. Molecular Modeling and Simulation. Springer; 2002.
9. Case, D.A., Cheatham, T.E., Darden, T., Gohlke, H., Luo, R., Merz, K.M., Onufriev, A., Simmerling, C., Wang, B., Woods, R.J. The amber biomolecular simulation programs. J. Comput. Chem. 2005, 26(16), 1668–88, December.
10. Gallicchio, E., Levy, R.M. Agbnp: An analytic implicit solvent model suitable for molecular dynamics simulations and high-resolution modeling. J. Comput. Chem. 2004, 25, 479–99.
11. Wagoner, J.A., Baker, N.A. Assessing implicit models for nonpolar mean solvation forces: The importance of dispersion and volume terms. Proc. Natl. Acad. Sci. USA 2006, 103(22), 8331–6, May.
12. Simonson, T. Electrostatics and dynamics of proteins. Rep. Prog. Phys. 2003, 66, 737–87.
13. Baker, N.A. Improving implicit solvent simulations: A Poisson-centric view. Curr. Opin. Struct. Biol. 2005, 15(2), 137–43, April.
14. Totrov, M., Abagyan, R. Rapid boundary element solvation electrostatics calculations in folding simulations: Successful folding of a 23-residue peptide. Biopolymers 2001, 60(2), 124–33.
15. Grant, A.J., Pickup, B.T., Nicholls, A. A smooth permittivity function for Poisson–Boltzmann solvation methods. J. Comput. Chem. 2001, 22(6), 608–40.
16. Prabhu, N.V., Zhu, P., Sharp, K.A. Implementation and testing of stable, fast implicit solvation in molecular dynamics using the smooth-permittivity finite difference Poisson–Boltzmann method. J. Comput. Chem. 2004, 25(16), 2049–64, December.
17. David, L., Luo, R., Gilson, M.K. Comparison of Generalized Born and Poisson models: Energetics and dynamics of HIV protease. J. Comput. Chem. 2000, 21(4), 295–309, March.
18. Bashford, D., Case, D. Generalized Born models of macromolecular solvation effects. Annu. Rev. Phys. Chem. 2000, 51, 129–52.
19. Dominy, B.N., Brooks, C.L. Development of a Generalized Born model parametrization for proteins and nucleic acids. J. Phys. Chem. B 1999, 103, 3765–73.
20. Calimet, N., Schaefer, M., Simonson, T. Protein molecular dynamics with the Generalized Born/ACE solvent model. Proteins 2001, 45, 144–58.
21. Tsui, V., Case, D. Molecular dynamics simulations of nucleic acids using a Generalized Born solvation model. J. Am. Chem. Soc. 2000, 122, 2489–98.
22. Wang, T., Wade, R. Implicit solvent models for flexible protein–protein docking by molecular dynamics simulation. Proteins 2003, 50, 158–69.
23. Nymeyer, H., Garcia, A.E. Free in pmc simulation of the folding equilibrium of alpha-helical peptides: A comparison of the Generalized Born approximation with explicit solvent. Proc. Natl. Acad. Sci. USA 2003, 100, 13934–49.
24. Still, W.C., Tempczyk, A., Hawley, R.C., Hendrickson, T. Semianalytical treatment of solvation for molecular mechanics and dynamics. J. Am. Chem. Soc. 1990, 112, 6127–9.
25. Srinivasan, J., Trevathan, M., Beroza, P., Case, D. Application of a pairwise Generalized Born model to proteins and nucleic acids: Inclusion of salt effects. Theor. Chem. Acts 1999, 101, 426–34.
26. Hawkins, G.D., Cramer, C.J., Truhlar, D.G. Pairwise solute descreening of solute charges from a dielectric medium. Chem. Phys. Lett. 1995, 246, 122–9.
27. Hawkins, G.D., Cramer, C.J., Truhlar, D.G. Parametrized models of aqueous free energies of solvation based on pairwise descreening of solute atomic charges from a dielectric medium. J. Phys. Chem. 1996, 100, 19824–36.
28. Scarsi, M., Apostolakis, J., Caflisch, A. Continuum electrostatic energies of macromolecules in aqueous solutions. J. Phys. Chem. A 1997, 101(43), 8098–106, October.
29. Onufriev, A., Bashford, D., Case, D.A. Modification of the Generalized Born model suitable for macromolecules. J. Phys. Chem. B 2000, 104(15), 3712–20, April.

30. Lee, M.S., Salsbury, F.R., Brooks III, C.L. Novel Generalized Born methods. J. Chem. Phys. 2002, 116, 10606–14.
31. Lee, M.S., Feig, M., Salsbury, F.R., Brooks, C.L. New analytic approximation to the standard molecular volume definition and its application to Generalized Born calculations. J. Comput. Chem. 2003, 24(11), 1348–56, August.
32. Onufriev, A., Bashford, D., Case, D.A. Exploring protein native states and large-scale conformational changes with a modified Generalized Born model. Proteins 2004, 55(2), 383–94, May.
33. Ghosh, A., Rapp, C.S., Friesner, R.A. Generalized Born model based on a surface integral formulation. J. Phys. Chem. B 1998, 102, 10983–90.
34. Onufriev, A., Case, D., Bashford, D. Effective Born radii in the Generalized Born approximation: The importance of being perfect. J. Comput. Chem. 2002, 23, 1297–304.
35. Mongan, J., Svrcek-Seiler, W.A., Onufriev, A. Analysis of integral expressions for effective Born radii. J. Chem. Phys. 2007, 127(18), 18510–1, November.
36. Grycuk, T. Deficiency of the Coulomb-field approximation in the Generalized Born model: An improved formula for Born radii evaluation. J. Chem. Phys. 2003, 119(9), 4817–26, September.
37. Swanson, J.M.J., Mongan, J., McCammon, J.A. Limitations of atom-centered dielectric functions in implicit solvent models. J. Phys. Chem. B 2005, 109(31), 14769–72, August.
38. Feig, M., Onufriev, A., Lee, M.S., Im, W., Case, D.A., Brooks, C.L. Performance comparison of Generalized Born and Poisson methods in the calculation of electrostatic solvation energies for protein structures. J. Comput. Chem. 2004, 25, 265–84.
39. Mongan, J., Simmerling, C., McCammon, J., Case, D., Onufriev, A. Generalized Born model with a simple, robust molecular volume correction. J. Chem. Theory Comput. 2007, 3, 156–69.
40. Simmerling, C., Strockbine, B., Roitberg, A.E. All-atom structure prediction and folding simulations of a stable protein. J. Am. Chem. Soc. 2002, 124, 11258–9.
41. Zagrovic, B., Snow, C.D., Shirts, M.R., Pande, V.S. Simulation of folding of a small alpha-helical protein in atomistic detail using worldwide-distributed computing. J. Mol. Biol. 2002, 323(5), 927–37, November.
42. Jang, S., Kim, E., Shin, S., Pak, Y. Ab initio folding of helix bundle proteins using molecular dynamics simulations. J. Am. Chem. Soc. 2003, 125(48), 14841–6, December.
43. Pitera, J.W., Swope, W. Understanding folding and design: Replica-exchange simulations of "trp-cage" miniproteins. Proc. Natl. Acad. Sci. USA 2003, 100(13), 7587–92, June.
44. Hornak, V., Okur, A., Rizzo, R.C., Simmerling, C. Hiv-1 protease flaps spontaneously open and reclose in molecular dynamics simulations. Proc. Natl. Acad. Sci. USA 2006, 103(4), 915–20, January.
45. Tanizaki, S., Feig, M. A Generalized Born formalism for heterogeneous dielectric environments: Application to the implicit modeling of biological membranes. J. Chem. Phys. 2005, 122(12), 12470–6, March.
46. Im, W., Brooks, C.L. Interfacial folding and membrane insertion of designed peptides studied by molecular dynamics simulations. Proc. Natl. Acad. Sci. USA 2005, 102(19), 6771–6, May.
47. Spassov, V.Z., Yan, L., Szalma, S. Introducing an implicit membrane in Generalized Born/solvent accessibility continuum solvent models. J. Phys. Chem. B 2002, 106, 8726–38.
48. Efremov, R.G., Nolde, D.E., Konshina, A.G., Syrtcev, N.P., Arseniev, A.S. Peptides and proteins in membranes: What can we learn via computer simulations? Curr. Med. Chem. 2004, 11(18), 2421–42, September.
49. Tsui, V., Case, D. Theory and applications of the Generalized Born solvation model in macromolecular simulations. Biopolymers 2001, 56, 275–91.
50. Sorin, E., Rhee, Y., Nakatani, B., Pande, V. Insights into nucleic acid conformational dynamics from massively parallel stochastic simulations. Biophys. J. 2003, 85(2), 790–803, August. Force field: AMBER95 Allen's stochastic (Langevin) integrator TINKER package Folding@Home GB/SA (Qui et al., 1997) TINKER.
51. De Castro, L.F., Zacharias, M. DAPI binding to the DNA minor groove: A continuum solvent analysis. J. Mol. Recognit. 2002, 15(4), 209–20, July–August.
52. Allawi, H., Kaiser, M., Onufriev, A., Ma, W., Brogaard, A., Case, D., Neri, B., Lyamichev, V. Modeling of flap endonuclease interactions with DNA substrate. J. Mol. Biol. 2003, 328, 537–54.
53. Chocholousová, J., Feig, M. Implicit solvent simulations of DNA and DNA–protein complexes: Agreement with explicit solvent vs experiment. J. Phys. Chem. B 2006, 110(34), 17240–51, August.

54. Ruscio, J.Z., Onufriev, A. A computational study of nucleosomal DNA flexibility. Biophys. J. 2006, 91(11), 4121–32, December.

55. Lee, M.S., Salsbury, F.R., Brooks, C.L. Constant-ph molecular dynamics using continuous titration coordinates. Proteins 2004, 56(4), 738–52, September.

56. Mongan, J., Case, D.A., McCammon, J.A. Constant pH molecular dynamics in Generalized Born implicit solvent. J. Comput. Chem. 2004, 25(16), 2038–48, December.

57. Mobley, D.L., Ii, A.E., Fennell, C.J., Dill, K.A. Charge asymmetries in hydration of polar solutes. J. Phys. Chem. B 2008, 112(8), 2405–14, February.

58. Sigalov, G., Scheffel, P., Onufriev, A. Incorporating variable dielectric environments into the Generalized Born model. J. Chem. Phys. 2005, 122(9), 9451, March.

59. Zhou, R., Berne, B.J. Can a continuum solvent model reproduce the free energy landscape of a beta-hairpin folding in water?, Proc. Natl. Acad. Sci USA 2002, 99(20), 12777–82, October.

60. Zhou, R., Krilov, G., Berne, B.J. Comment on 'can a continuum solvent model reproduce the free energy landscape of a beta-hairpin folding in water?' The Poisson–Boltzmann equation. J. Phys. Chem. B 2004, 108, 7528–30.

61. Roe, D.R., Okur, A., Wickstrom, L., Hornak, V., Simmerling, C. Secondary structure bias in Generalized Born solvent models: Comparison of conformational ensembles and free energy of solvent polarization from explicit and implicit solvation. J. Phys. Chem. B 2007, 111(7), 1846–57, February.

62. Ramstein, J., Lavery, R. Energetic coupling between DNA bending and base pair opening. Proc. Natl. Acad. Sci. USA 1988, 85(19), 7231–5, October.

63. Wang, L., Hingerty, B.E., Srinivasan, A.R., Olson, W.K., Broyde, S. Accurate representation of B-DNA double helical structure with implicit solvent and counterions. Biophys. J. 2002, 83(1), 382–406, July.

64. Kosikov, K.M., Gorin, A.A., Lu, X.J., Olson, W.K., Manning, G.S. Bending of DNA by asymmetric charge neutralization: All-atom energy simulations. J. Am. Chem. Soc. 2002, 124(17), 4838–47, May.

65. Tironi, I.G., Sperb, R., Smith, P.E., van Gunsteren, W.F. A generalized reaction field method for molecular dynamics simulations. J. Chem. Phys. 1995, 102(13), 5451–9.

66. Sigalov, G., Fenley, A., Onufriev, A. Analytical electrostatics for biomolecules: Beyond the Generalized Born approximation. J. Chem. Phys. 2006, 124(12), 12490–2, March.

67. Abagyan, R., Totrov, M. Biased probability Monte Carlo conformational searches and electrostatic calculations for peptides and proteins. J. Mol. Biol. 1994, 235(3), 983–1002, January.

68. Cai, W., Deng, S., Jacobs, D. Extending the fast multipole method to charges inside or outside a dielectric sphere. J. Comput. Phys. 2006, 223, 846–64.

69. Rottler, J. Local electrostatics algorithm for classical molecular dynamics simulations. J. Chem. Phys. 2007, 127(13), 13410–4, October.

70. Lee, M.S., Salsbury, F.R., Olson, M.A. An efficient hybrid explicit/implicit solvent method for biomolecular simulations. J. Comput. Chem. 2004, 25(16), 1967–78, December.

71. Okur, A., Wickstrom, L., Layten, M., Geney, R., Song, K., Hornak, V., Simmerling, C. Improved efficiency of replica exchange simulations through use of a hybrid explicit/implicit solvation model. J. Chem. Theory Comput. 2006, 2(2), 420–33.

72. Prabhu, N.V., Panda, M., Yang, Q., Sharp, K.A. Explicit ion, implicit water solvation for molecular dynamics of nucleic acids and highly charged molecules. J. Comput. Chem. 2007, December.

73. Lazaridis, T., Karplus, M. Effective energy function for proteins in solution. Proteins 1999, 35(2), 133–52, May.

74. Dzubiella, J., Swanson, J.M., McCammon, J.A. Coupling nonpolar and polar solvation free energies in implicit solvent models. J. Chem. Phys. 2006, 124(8), February.

CHAPTER 8

Comparing MD Simulations and NMR Relaxation Parameters

Vance Wong* and **David A. Case***

Contents		

1. INTRODUCTION

Molecular dynamics (MD) simulations of proteins are now about 30 years old [1]. They have become increasingly common and useful as force fields have improved (providing more realistic descriptions of microscopic forces) and as computers have become more powerful (allowing longer simulations that explore more of the available conformational space). It was clear from quite early times that simulations could be useful in interpreting NMR experiments, both for fast motions seen primarily in NMR relaxation [2–4], and for the much slower motions relevant to "chemical exchange" [5]. Subsequent development of sophisticated experimental methods for following spin relaxation, particularly in isotopically-labeled samples, has led to renewed interest in exploring dynamical connections between simulation and experiment.

The fundamentals of NMR relaxation theory have been presented in many places [6–9], and there is no space here to give more than a taste of what is involved. The rate of return of a spin system to equilibrium is determined by the time-dependent magnetic fields experienced at each atomic nucleus, arising from molecular motions. The ability of this stochastic, fluctuating field to induce spin

* Department of Molecular Biology, The Scripps Research Institute, La Jolla, CA 92037, USA

Annual Reports in Computational Chemistry, Vol. 4
ISSN 1574-1400, DOI: 10.1016/S1574-1400(08)00008-X

transitions depends upon its intensity at frequencies that correspond to sums and differences of the Larmor frequencies of the nuclear spins. These are represented as "spectral densities" $J(\omega)$, which in turn are the Fourier transforms of microscopic time correlation functions. For example, the ability of an amide proton to relax the spin of the ^{15}N nucleus to which it is attached can be expressed in terms of the time correlation function

$$C(\tau) \equiv \langle P_2[\mu(t) \cdot \mu(t + \tau)] \rangle \tag{1}$$

where $\mu(t)$ is the (time-dependent) unit vector from the nitrogen to the proton, $P_2(\cos\theta) \equiv (3\cos^2\theta - 1)/2$ is a Legendre polynomial, and the brackets indicate a time- and ensemble-average over all conformations of the system. The spectral density $J(\omega)$ is then the Fourier transform of $C(\tau)$. To be effective in relaxation, $J(\omega)$ needs to be large near the Larmor frequency of the spins or near zero frequency. As it happens, the most effective molecular motion for biomolecules (from the standpoint of NMR spin relaxation) is overall rotational diffusion or tumbling. It is this fact that makes NMR structure determination possible in the first place: to a good first approximation, proton relaxation (as monitored by nuclear Overhauser effect measurements) can be interpreted as arising from the rotational Brownian motion of a *rigid* molecule, and the atomic coordinates of this hypothetical rigid structure can be adjusted to optimize agreement with the NMR data. It is important to note that this does not mean that internal deformations are not present, but only that most such motions are fairly inefficient in driving NMR spin relaxation. In general, a more careful quantitative analysis (usually involving measurements of "heteronuclear" ^{13}C or ^{15}N relaxation) is required to extract information about internal motions from NMR data.

In principle, the evaluation of time-correlation functions like that in Eq. (1) should provide a powerful tool to test the quality of MD simulations. In NMR experiments, the proteins are in thermal equilibrium (only the spins are out of equilibrium, and their energies are negligible), and there are (non-invasive) probes throughout the structure at both backbone and side-chain positions. This facilitates details and direct comparisons between liquid simulations and experiment. Furthermore, rotational tumbling itself drives the drives the time-correlation function to zero, so that motions on a time scale slower than rotational tumbling are not relevant for this sort of relaxation. Rotational diffusion times for soluble proteins are on the order of 10 ns for many systems, and it is now feasible to readily carry out simulations to this time scale and beyond. In the most favorable cases, which are small, compact and very stable proteins, we can hope to reduce sampling errors to an acceptable range, and to ascribe differences between simulation and experiment to biases in the force fields we are using, and to use such comparisons to help improve the physical realism of MD simulations.

This general subject has been covered in earlier reviews [10,11]. Here, we summarize recent (and significant) progress made possible by faster computers and longer simulations, and by improvements in protein force fields.

2. INTERNAL MOTIONS AND FLEXIBILITY

In the past, comparisons between NMR relaxation and MD simulations have concentrated on internal motions, since these often involve sub-nanosecond time scales that could be examined with limited computer resources. In this approach, overall rotational motion is removed by an rms fitting procedure (for example, on backbone atoms in regular secondary structure), and computing time-correlation functions from the result. Typical results are shown in the upper panel of Figure 8.1; similar plots have been presented many times before [4,12,10,11]. Many backbone vectors are like Thr 49, and decay in less than 0.1 ns to a plateau value which can be identified as the order parameter S^2 for that vector. Most regions of regular secondary structure resemble this, although there can be exceptions, and there is potentially important information in the decay rates and plateau values that are obtained.

Figure 8.1 also shows two examples of floppier residues, with order parameters less than 0.8 and internal decay times that are comparable to overall tumbling times, such as residue 41 in GB3, whose internal correlation function decays with a τ_e of 2.1 ns. Such slow decays have received less attention, since they can only be reliably observed with fairly long simulations: as a rule of thumb, simulation times need to be 50–100 times longer than the examined decay times in order to obtain reasonably converged time correlation functions [16,17,13]. With 100 ns simulations now becoming available (and which are needed to examine overall rotational diffusion anyway), better comparisons to experiment for these slower internal motions are becoming feasible.

Until quite recently, comparisons between calculated and measured order parameters for backbone NH vectors in proteins had decidedly mixed results. Some comparisons seemed very promising [18], but many calculations gave results more like the dotted lines in the lower panel of Figure 8.1. Here, only a handful of residues show NMR order parameters below 0.8, but many calculated values are below this value; basically, the protein exhibits more backbone mobility in the simulation than appears to be warranted. Similar behavior has been noted in many other simulations [19–21]. Recently, both the CHARMM and Amber force fields were modified by changes in the energetics of torsion terms for the ϕ and ψ backbone angles in proteins [22–24,20]. Although these changes were motivated by consideration of short peptides, they have had the effect of considerably reducing the amount of backbone motion in MD simulations, and systematically improving the agreement between calculated and observed order parameters [19–21]. Figure 8.2 shows some new comparisons that complement those already published. The upper panel compares two simulations using the Amber ff99SB force field, carried out in different labs. One was 30 ns in length and the other was 100 ns. The two simulations used similar but not identical water models (TIP3P vs. SPC/E), so an identical match of results should not be expected. Nevertheless, there is a close correspondence between the two data sets, showing that the water model seems to have little effect on order parameters, and that independent simulations can give very similar results. The largest discrepancy is for residue 69 in lysozyme (marked in the figure); here we suspect that the longer simulation has a better converged

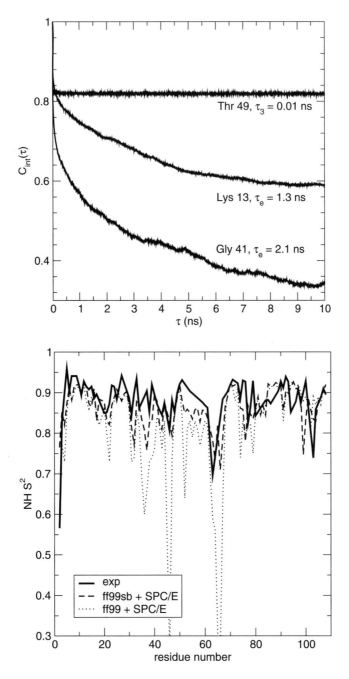

FIGURE 8.1   Upper: Internal correlation functions for NH vectors in selected residues in a 200 ns simulation of protein GB3 + SPC/E water [13]. Values of τ_e show the model-free value for the decay time of these internal correlation functions [14]. Lower: Order parameters for NH vectors in binase; experimental values from [15], computed values from simulations described in [13].

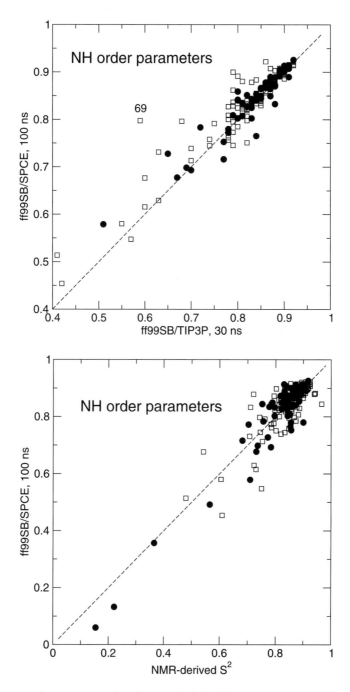

FIGURE 8.2 NH order parameters for ubiquitin (solid circles) and lysozyme (open squares). Upper: comparison of MD results from [20] to those computed here from data in [13]. Lower: values derived from NMR compared to the MD simulations of [13].

correlation function which (as it happens) is in better agreement with the NMR value of 0.74.

The lower panel of Figure 8.2 compares NMR and simulated order parameters for the same two proteins. This shows a linear correlation coefficient R of 0.91, and a root-mean-square difference between the two data sets of 0.05 over a wide range of motional amplitudes. Hornak et al. [20] present evidence suggesting that improved order parameters arise from residues in certain regions of backbone (ϕ, ψ) space, which were in somewhat unfavorable regions in previous force fields, which led to structural fluctuations as the backbone terms conflicted with the topology and side-chain packing that led to that structure in the first place. Relieving this backbone "strain" removed some of the impetus for the larger amplitude motions. Overall, Figures 8.1 and 8.2 (and comparable data in the papers cited here) provide a snapshot of the performance of current simulations as monitored by backbone motional amplitudes.

Showalter et al. [25] have recently extended this sort of analysis to methyl side-chain dynamics with encouraging results, including a similar marked improvement when using recent force fields (Amber ff99SB) compared to earlier ones (Amber ff99). The performance of MD simulations for side-chain conformations can also be compared to experiment for a few small proteins where populations of side chain conformers can be estimated from NMR data. Fragment B3 of protein G (see the cartoon at the upper part of Figure 8.3) is an especially favorable case where both heteronuclear coupling constants and residual dipolar coupling information has been combined to obtain population estimates for side-chain rotamers for Val, Ile, and Thr residues [26]. The lower part of Figure 8.3 compares rotamer populations (g^+, g^- and t) found in a 100 ns simulation (using the Amber ff99SB force field and the TIP4P/EW water model) with those estimated from NMR. There is clearly a lot of scatter in the comparison, but also a clear division between predominant conformers (with populations greater than 0.5) and minor conformers (with populations less than 0.5). In each of the 16 side-chains analyzed, the predominant conformer is the same in experiment and theory. Furthermore, three of the 16 residues (Val 42, Thr 11 and Thr 55) started in a conformer (taken from the X-ray structure) that is different from the most highly populated conformer determined by the NMR analysis, so that the simulation was able to sample and favor conformers other than those in the starting structure.

There are two respects in which the discussion so far has been oversimplified. First, zero-point vibrational motion, a quantum effect that is ignored in classical simulations, can have a small but noticeable effect on internal order parameters [27,28]. Second, the analysis of NMR relaxation data uses as an input the N–H distance, which is often treated as an empirical parameter that is slightly larger than the true average distance. These two features are closely related to each other: the use of an effective distance slightly larger than the true one represents an attempt to include the effects of local peptide zero-point motions, so that the remaining motion (represented by the order parameter) would represent "interesting" disorder that is beyond that what would be found even in a frozen peptide at 0 K [28].

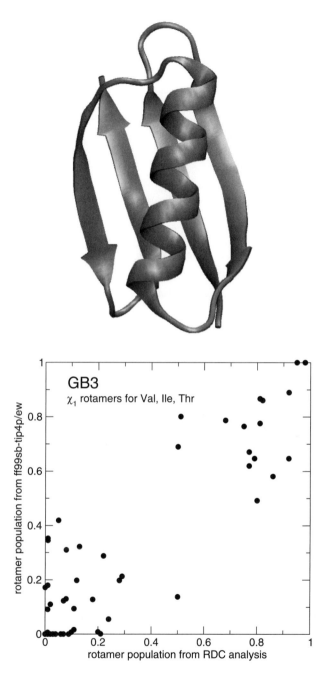

FIGURE 8.3   Upper: Backbone cartoon for GB3. Lower: Comparison of rotamer populations estimated from NMR data [26] with those computed from the ff99SB, TIP4P/EW trajectory reported in [13].

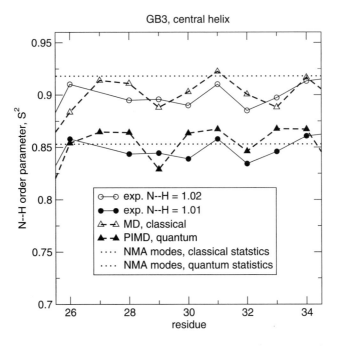

FIGURE 8.4 Quantum effects on the order parameters for the NH vectors in the central helix of GB3.

Figure 8.4 illustrates some of these features for the central helix of GB3. The top three curves (centered around 0.9) show "classical" behavior, whereas the bottom three curves (centered around 0.85) illustrate quantum behavior. The classical calculations, either from standard MD simulations or from a normal mode analysis of NMA where zero-point motion is not enforced, are compared to results from experiment where the NH bond is assumed to be slightly longer than its true value of 1.01 Å. The quantum calculations are based on path-integral molecular dynamics calculations [29] or from NMA normal modes that enforce quantum statistics (and hence zero-point motion); these are compared to an analysis of experiment using a more correct estimate of the NH bond length. It is encouraging that the quantum calculations match closely the results extrapolated from NMR with a realistic bond length. Furthermore, the fact that local normal mode motion (the dotted line with a value slightly above 0.85) comes close to matching experiment is illuminating. This implies that a model in which the protein is completely rigid, except for (inevitable) local zero-point motions, gives enough motion to explain the relaxation data for this piece of secondary structure, and that no additional internal fluctuations of the helix are needed to explain the data. This suggests that many pieces of secondary structure in well-folded proteins are about as rigid as they could possibly be [30,28], and that earlier simulations (or interpretations of experimental order parameters) have overstated to some extent the amount of internal motion in such secondary structure.

3. OVERALL TUMBLING AND ROTATIONAL DIFFUSION

Overall rotational tumbling is regulated by frequent collisions with light water molecules. For a nearly rigid protein, this physical model should lead to diffusive rotational behavior, where the reorientation of a unit vector attached to the molecule undergoes a random walk on the surface a sphere. If $c(\mathbf{n}, t)$ is the probability density for finding the vector pointing direction $\mathbf{n}$ at time t, a spherical molecule should follow a simple diffusion equation [31,32]:

$$\frac{\partial c(\mathbf{n}, t)}{\partial t} = D_{\text{rot}} \nabla^2 c(\mathbf{n}, t) = -D_{\text{rot}} \hat{l}^2 c(\mathbf{n}, t). \tag{2}$$

Here $\hat{l}$ is a (dimensionless) angular momentum operator. A non-spherical molecule will tumble more rapidly about some directions than about others, causing the diffusion constant D_{rot} to become a tensor:

$$\frac{\partial c(\Omega, t)}{\partial t} = -\sum_{i,j} \hat{l}_i \cdot \mathbf{D}_{ij} \cdot \hat{l}_j c(\Omega, t). \tag{3}$$

Here Ω represents the Euler angles that specify the orientation of the macromolecule.

The most powerful way to measure macromolecular diffusion is by NMR relaxation, since it is very sensitive to both the overall tumbling frequency and its anisotropy [33,9]. The analysis of NMR relaxation data typically *assumes* that the rotational motion of a compact and folded protein follows Eq. (3), so that the goal of the analysis is to determine the principal values and orientation of the diffusion tensor $\mathbf{D}$. Deviations from the behavior predicted for a single diffusion tensor are generally taken as evidence for internal motion (i.e. for non-rigid behavior), most commonly using a model-free formalism that assumes a statistical independence of internal and overall motion [14,11]. While there is no question that this overall description is qualitatively correct for many well-folded proteins, quantitative analyses of NMR relaxation data increasingly face questions about the correctness of these assumptions. Does overall rotation follow diffusion theory (with a single tensor $\mathbf{D}$), or would it more correct to adopt a model with a distribution of tensors or correlation times [14,34]? Are typical internal motions of proteins uncorrelated with overall rotational motion, and can we approximate the full correlation function as a product of separate overall and internal functions? As proteins become more disordered (either *in toto*, or as a result of floppy "tails") how quickly to these standard models fail?

In principle, molecular dynamics simulations should have a lot to say about these questions, since they provide a detailed (albeit approximate) description of macromolecular structure and dynamics. One can learn some information about global motion by extrapolations from even short simulations [39], but the longer time scales now available, which can be many times the mean rotational tumbling time, are expected to yield more reliable information. However, many popular water models (such as TIP3P) predict self-diffusion constants (and, presumably, viscosities) that are far from experiment (see Table 8.1), so that one would not

Table 8.1 Self-diffusion constants for the models of water used here. Data for TIP4P/EW from [35]; Amoeba from [36]; SPC/Fw and q-SPC/Fw from [37]; remaining data from [38]

Model	$D^{298}, 10^{-9}\,\mathrm{m^2\,s^{-1}}$
Experiment	2.2
TIP3P	5.7
SPC/E	2.8
TIP4P/EW	2.3
TIP5P	2.6
Amoeba	2.0
SPC/Fw	2.3
q-SPC/Fw	2.4

expect good results for rotational or translation diffusion of macromolecules dissolved in such solvents. Finally, many biomolecular simulations use thermostats or barostats that can affect dynamical properties in ways that are not well understood. We have recently started a re-examination of this problem, using 100 to 200 ns simulations of some small, well-folded proteins [13].

For isotropic diffusional motion, solutions to Eq. (2) are easily computed for any value of ℓ, and the correlation functions are single exponentials [32]:

$$C_\ell(\tau) = \exp\left[-\ell(\ell+1)D_{\mathrm{rot}}\tau\right]. \tag{4}$$

Things are more complex for anisotropic molecules, even though the general solution of the anisotropic rigid body diffusion problem has been known for many years [31]. It is a straightforward but algebraically complex matter to compute the time correlation function for a vector fixed to the rigid body using the Green's function of the rotational diffusion operator [40,8]. The same result has been obtained without direct use of the diffusion operator eigenfunctions [41]. Rank 2 correlation functions have five exponentials that have be written down in many places[42,43,41,32,44]. For simplicity, we show here results for a symmetric top (where $D_x = D_y$), which has three exponential terms

$$C_2(\tau) = \frac{1}{4}\Big\{(3\cos^2\theta - 1)^2 \exp[-6D_x\tau]$$

$$+ 12\cos^2\theta \sin^2\theta \exp\left[-(5D_x + D_z)\tau\right]$$

$$+ 3\sin^4\theta \exp\left[-(2D_x + 4D_z)\tau\right]\Big\}. \tag{5}$$

Hence, for anisotropic rotation, the simple dependence of the decay times on $\ell(\ell+1)$ seen in Eq. (4) no longer holds, although it is still on average valid to first order in the anisotropy (see Eq. (8) below).

3.1 Fitting diffusion tensors to MD data

In order to determine the diffusion tensor that best fits a given trajectory, we use a procedure that was developed for the analysis of NMR relaxation data [45–47], and which is well-suited for adaptation to the analysis of MD simulations. This uses the average (or "effective") correlation time for a particular vector, $\mathbf{n}$ (e.g. a backbone N–H bond vector, or a randomly chosen direction in the molecular frame), which can be defined as

$$\tau_\ell(\mathbf{n}) = \int_0^\infty d\tau \langle P_\ell[\mathbf{n}(0) \cdot \mathbf{n}(\tau)] \rangle \tag{6}$$

where

$$\langle P_\ell[\mathbf{n}(0) \cdot \mathbf{n}(\tau)] \rangle = \lim_{T \to \infty} \frac{1}{T} \int_0^T P_\ell[\mathbf{n}(t) \cdot \mathbf{n}(t - \tau)] \, dt \tag{7}$$

is the usual time correlation function of a Legendre polynomial of order ℓ. Note that $\tau(\mathbf{n}) \propto J(0)$, the zero-frequency component of the corresponding spectral density function. This correlation time is related to a "local" or effective diffusion constant by:

$$d_{\text{loc}}(\mathbf{n}, \ell) \equiv \frac{1}{\ell(\ell + 1)\tau_\ell(\mathbf{n})}. \tag{8}$$

For diffusion tensors with small anisotropy, $d_{\text{loc}}(\mathbf{n}, \ell)$ may be written as a quadratic function in $\mathbf{n}$ [45–47]:

$$d_{\text{loc}}(\mathbf{n}, \ell) = \mathbf{n}^T \cdot \mathbf{Q} \cdot \mathbf{n} \tag{9}$$

where

$$\mathbf{Q} = \frac{3D_{\text{av}}\mathbf{I} - \mathbf{D}}{2}. \tag{10}$$

Since the right-hand side of Eq. (9) is independent of ℓ, d_{loc} should also be independent of ℓ when the motion is well characterized as rotational diffusion of $\mathbf{n}$. Indeed, one test of whether a diffusion model fits the MD data is to examine the dependence of d_{loc} (or, equivalently τ_ℓ) on ℓ [32,39].

In NMR experiments, $\mathbf{n}$ is typically a backbone N–H bond vector, and $d_{\text{loc}}(\mathbf{n}, 2)$ is computed as a function of $R_2(\mathbf{n})/R_1(\mathbf{n})$ (or related quantities, such as $(2R_2 - R_1)/R_1$), where R_1 and R_2 are longitudinal and transverse relaxation rates [48,47]. In this way, the local diffusion constants become a key intermediate quantity that can be estimated from both NMR experiments and from simulations; in this respect, they play much the same role here as the model-free parameters S^2 and τ_e play in the analysis of internal motions by MD.

From the simulation data, the correlation time is found by integrating the time correlation function as shown in Eq. (6). While the correlation functions may be easily computed from the trajectories, statistical errors due to finite trajectory length limit the useful data to short delay times τ [16,17]. The variance of a Gaus-

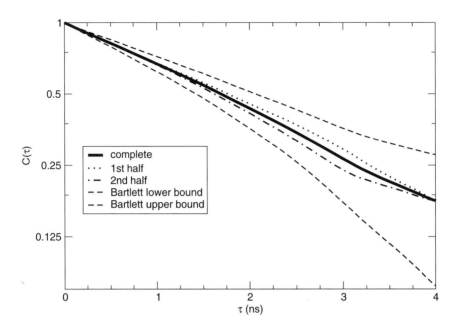

FIGURE 8.5   Semi-log plot of $C_2(\tau)$ vs. delay time, for a randomly-chosen direction in the GB3 + SPC/E simulation (solid black line), with values computed from pieces of the trajectory in dot and dash-dot lines. Dashed lines are the error bounds from Eq. (11), using $T = 200$ ns, $\tau = 2.5$ ns, so that $T_s = 80$.

sian process may be estimated using the Bartlett formula [49,50], which when applied to an isotropic rotational diffusion process yields [16,17]:

$$\sigma(j) = \left\{ \frac{1}{T_s} [1 - (1 + 2j_s)\exp(-2j_s)] \right\}^{1/2}. \tag{11}$$

Here, j indexes the data points in the time series, $T_s = T/\tau$ is the scaled trajectory length, and $j_s \equiv j/\tau$ is the scaled delay. Figure 8.5 shows a single $\ell = 2$ correlation function generated from a member of the random initial vector set for a 200 ns simulation of GB3, together with estimates of variance computed using Eq. (11), and a very simple uncertainty analysis that just used either the first or the second half of the trajectory. The expected variance is small for short delays, but becomes larger as $C(\tau)$ decays. For the parameters of the figure, the expected uncertainty in $C(\tau)$ is about 0.06 at $j_s = 1/2$, and grows to 0.09 by the time $j_s = 1$, that is, at a delay time equal to the rotational correlation time; the latter value corresponds to a 23% relative uncertainty, which clearly has a significant impact on the ability to estimate the rate at which the true correlation function decays. However, the analysis of partial trajectories (shown in more detail in [13]) suggest that the uncertainties estimated by Eq. (11) are wider than they need to be. In any event, it is clear that long trajectories (of hundreds of nanoseconds) may be required to obtain good precision in the fitted diffusion tensors.

3.2 Do the simulations exhibit diffusive motion?

While it is always possible to find the best-fit diffusion tensor from the correlation functions of a reorienting molecule, the rotational dynamics need not be diffusive. For example, the rotation of small molecules in solution can be much more inertial in character than one would ordinarily expect for a macromolecule [32]. A more relevant scenario to protein dynamics would be instances where a single diffusion tensor cannot describe global tumbling because of conformational transitions that change the shape of the protein. However, for the small, well-folded proteins considered in [13] (GB3, ubiquitin, binase and lysozyme), the overall rotation was well characterized as diffusion, as monitored by the comparisons between MD and diffusional estimates of $d_{loc}(\mathbf{n}, l)$ for $l = 1$ to 8.

Aside from the TIP3P trajectories, the simulation values of D_{av} are 10 to 30 percent larger than their experimental counterparts, as might be expected from the self-diffusion constants given in Table 8.1. One can also look at the anisotropy of tumbling, as compared to NMR results. As an example, Figure 8.6 shows d_{loc} for NH bonds as a function of residue for GB3. The "comb-like structure" (between residues 23 and 39) in the d_{loc} data for GB3 is characterized by slower diffusion than the remainder of the protein. This is due to alignment of NH bonds within the central helix along the helix axis, which is also roughly parallel to the diffusion tensor axis of symmetry [51], as can be seen in Figure 8.3. Overall, the shapes of the three curves are in rough agreement with one another, but the degree of contrast between the central α-helix (residues 25–35) and the surrounding β-sheets is greater in the simulations than the experiments. It is likely that this is related to

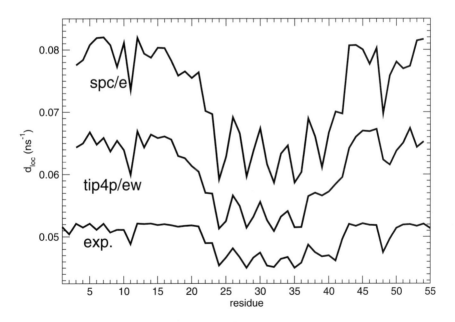

FIGURE 8.6 Local diffusion constants as a function of residue number for GB3; simulation date from [13], and experimental values are from [51].

the fact that overall tumbling is also faster in the simulations than in experiment. As proteins tumble, a rough "hydration shell" travels with them, increasing their effective size and slowing the motion [52]. This extra shell also tends to make the effective rotating entity more isotropic, so it is reasonable to speculate that current protein/water potentials slightly underestimate the association of water to the protein surface, leading to rotational diffusion tensors that are both too large and too anisotropic.

4. CONCLUSIONS

There are many ways in which one might monitor the fidelity of protein MD simulations to the underlying physical system. Analyses of structures and energies are common and powerful criteria that have long been used to test and improve force fields. NMR relaxation also provides a complementary view that emphasizes time scales and amplitudes of motional excursions about an average structure. The relevant time scales (bounded by rotational diffusion) are just now becoming routinely accessible to MD simulation, and it is likely that important information and constraints on simulation quality will continue to appear from these sorts of comparisons. Already, there is a noticeable trend away from TIP3P water, whose dynamical properties are quite far from experiment, in favor of models with stronger hydrogen-bonding interactions. In a similar fashion, one argument (among many) for the use of recent adjustments to torsional angle potentials lies in the reduced amplitude of fluctuations that are seen in simulations using these newer force fields, and analysis of NMR relaxation has been useful in confirming that such reductions are warranted. There is certainly much more to be done: for example, Figures 8.3 and 8.6 still show results with significant deviations from experiment. But results like those outlined here do set the "bar" higher for what we expect for simulated behavior of well-folded proteins in water.

ACKNOWLEDGMENTS

This work was supported by NIH grant GM45811. We thank David Fushman and Jennifer Hall for helpful discussions and for providing details of their experimental analyses.

REFERENCES

1. McCammon, J.A., Gelin, B.R., Karplus, M. Dynamics of folded proteins. Nature 1977, 267, 585–90.
2. Levy, R.M., Karplus, M., Wolynes, P.G. NMR relaxation parameters with internal motion: Exact Langevin trajectory results compared with simplified relaxation models. J. Am. Chem. Soc. 1981, 103, 5998–6011.
3. Levy, R.M., Dobson, C.M., Karplus, M. Dipolar NMR relaxation of nonprotonated aromatic carbons in proteins. Structural and dynamical effects. Biophys. J. 1982, 39, 107–13.
4. Lipari, G., Szabo, A., Levy, R.M. Protein dynamics and NMR relaxation: Comparison of simulations with experiment. Nature 1982, 300, 197–8.

5. Northrup, S.H., Pear, M.R., Lee, C.-Y., McCammon, J.A., Karplus, M. Dynamical theory of activated processes in globular proteins. Proc. Natl. Acad. Sci. USA 1982, 79, 4035–9.
6. Goldman, M. Quantum Description of High-Resolution NMR in Liquids. Oxford: Clarendon Press; 1988.
7. Fischer, M.W.F., Majumdar, A., Zuiderweg, E.R.P. Protein NMR relaxation: Theory, applications and outlook. Prog. Nucl. Magn. Reson. Spectrosc. 1998, 33, 207–72.
8. Korzhnev, D.M., Billeter, M., Arseniev, A.S., Orehkov, V.Y. NMR studies of tumbling and internal motions in proteins. Prog. NMR Spectr. 2001, 38, 197–266.
9. Cavanagh, J., Fairbrother, W.J., Palmer III, A.G., Rance, M., Skelton, N.J. Protein NMR Spectroscopy: Principles and Practice. second ed. Burlington, MA: Academic Press; 2007.
10. Case, D.A. Molecular dynamics and NMR spin relaxation in proteins. Acct. Chem. Res. 2002, 35, 325–31.
11. Palmer III, A.G. NMR characterization of the dynamics of biomacromolecules. Chem. Rev. 2004, 104, 3623–40.
12. Palmer, A.G., Case, D.A. Molecular dynamics analysis of NMR relaxation in a zinc-finger peptide. J. Am. Chem. Soc. 1992, 114, 9059–67.
13. Wong, V., Case, D.A. Evaluating rotational diffusion from protein MD simulations. J. Phys. Chem. B 2008, 112, 6013–24.
14. Lipari, G., Szabo, A. Model-free approach to the interpretation of nuclear magnetic resonance relaxation in macromolecules. I. Theory and range of validity. J. Am. Chem. Soc. 1982, 104, 4546–59.
15. Pang, Y., Buck, M., Zuiderweg, E.R.P. Backbone dynamics of the ribonuclease binase active site area using multinuclear (^{15}N and ^{13}CO) NMR relaxation and computational molecular dynamics. Biochemistry 2002, 41, 2655–66.
16. Zwanzig, R., Ailawadi, N.K. Statistical error due to finite time averaging in computer experiments. Phys. Rev. 1969, 182, 280–3.
17. Lu, C.Y., van den Bout, D.A. Effect of finite trajectory length on the correlation function analysis of single molecule data. J. Chem. Phys. 2006, 125, 1247–51.
18. Philippopoulos, M., Mandel, A.M., Palmer, A.G. III, Lim, C. Accuracy and precision of NMR relaxation experiments and MD simulations for characterizing protein dynamics. Proteins 1997, 28, 481–93.
19. Buck, M., BouguetBonnet, S., Pastor, R.W., MacKerell, A.D. Importance of the CMAP correction to the CHARMM22 protein force field: Dynamics of hen lysozyme. Biophys. J. 2006, 90, L36–8.
20. Hornak, V., Abel, R., Okur, A., Strockbine, B., Roitberg, A., Simmerling, C. Comparison of multiple Amber force fields and development of improved protein backbone parameters. Proteins 2006, 65, 712–25.
21. Showalter, S.A., Brüschweiler, R. Validation of molecular dynamics simulations of biomolecules using NMR spin relaxation as benchmarks: Application to the Amber99SB force field. J. Chem. Theory Comput. 2007, 3, 961–75.
22. MacKerell, A.D. Empirical force fields for biological macromolecules: Overview and issues. J. Comput. Chem. 2004, 25, 1584–604.
23. MacKerell, A.D., Feig, M., Brooks, C.L. Improved treatment of the protein backbone in empirical force fields. J. Am. Chem. Soc. 2004, 126, 698–9.
24. Duan, Y., Wu, C., Chowdhury, S., Lee, M.C., Xiong, G., Zhang, W., Yang, R., Cieplak, P., Luo, R., Lee, T. A point-charge force field for molecular mechanics simulations of proteins based on condensed-phase quantum mechanical calculations. J. Comput. Chem. 2003, 24, 1999–2012.
25. Showalter, S.A., Johnson, E., Rance, M., Brüschweiler, R. Toward quantitative interpretation of methyl side-chain dynamics from NMR by molecular dynamics simulations. J. Am. Chem. Soc. 2007, 129, 14146–7.
26. Chou, J.J., Case, D.A., Bax, A. Insights into the mobility of methyl-bearing side chains in proteins from 3JCC and 3JCN couplings. J. Am. Chem. Soc. 2003, 125, 8959–66.
27. Brüschweiler, R. Normal modes and NMR order parameters in proteins. J. Am. Chem. Soc. 1992, 114, 5341–4.
28. Case, D.A. Calculations of NMR dipolar coupling strengths in model peptides. J. Biomol. NMR 1999, 15, 95–102.
29. Tang, S., Case, D. Vibrational averaging of chemical shift anisotropies in model peptides. J. Biomol. NMR 2007, 38, 255–66.

30. Buck, M., Karplus, M. Internal and overall peptide group motion in proteins: Molecular dynamics simulations for lysozyme compared with results from X-ray and NMR spectroscopy. J. Am. Chem. Soc. 1999, 121, 9645–58.
31. Favro, L.D. Theory of the rotational Brownian motion of a free rigid body. Phys. Rev. 1960, 119, 53.
32. Berne, B.J., Pecora, R. Dynamic Light Scattering, with Applications to Chemistry, Biology and Physics. Mineola, NY: Dover; 2000.
33. Abragam, A. Principles of Nuclear Magnetis. Cambridge: Clarendon Press; 1961.
34. Ochsenbein, F., Neumann, J.M., Guittet, E., van Heijenoort, C. Dynamical characterization of residual and non-native structures in a partially folded protein by 15N NMR relaxation using a model based on a distribution of correlation times. Prot. Sci. 2002, 11, 957–64.
35. Horn, H.W., Swope, W.C., Pitera, J.W., Madura, J.D., Dick, T.J., Hura, G.L., Head-Gordon, T. Development of an improved four-site water model for biomolecular simulations: TIP4P-Ew. J. Chem. Phys. 2004, 120, 9665–78.
36. Ren, P., Ponder, J.W. Temperature and pressure dependence of the AMOEBA water model. J. Phys. Chem. B 2004, 108, 13427–37.
37. Paesani, F., Zhang, W., Case, D.A., Cheatham, T.E., Voth, G.A. An accurate and simple quantum model for liquid water. J. Chem. Phys. 2006, 125, 1845–7.
38. Mark, P., Nilsson, L. Structure and dynamics of the TIP3P, SPC, and SPC/E water models at 298 K. J. Phys. Chem. A 2001, 105, 9954–60.
39. Smith, P.E., van Gunsteren, W.F. Translational and rotational diffusion of proteins. J. Mol. Biol. 1994, 236, 629–36.
40. Huntress, W.T. Effects of anisotropic molecular rotational diffusion on nuclear magnetic relaxation in liquids. J. Chem. Phys. 1968, 48, 3524–33.
41. Woessner, D.E. Nuclear spin relaxation in ellipsoids undergoing rotational Brownian motion. J. Chem. Phys. 1962, 37, 647–54.
42. Hubbard, P.S. Nuclear magnetic relaxation of three and four spin molecules in a liquid. Phys. Rev. 1958, 109, 1153–8.
43. Woessner, D.E. Spin-relaxation processes in a two-proton system undergoing anisotropic reorientation. J. Chem. Phys. 1962, 36, 1–4.
44. Schwalbe, H., Carlomagno, T., Hennig, M., Junker, J., Reif, B., Richter, C., Griesinger, C. Cross-correlated relaxation for measurement of angles between tensorial interactions. Meth. Enzymol. 2001, 338, 35–81.
45. Brüschweiler, R., Liao, X., Wright, P.E. Long-range motional restrictions in a multidomain zinc-finger protein from anisotropic tumbling. Science 1995, 268, 886–9.
46. Lee, L.K., Rance, M., Chazin, W.J., Palmer III, A.G. Rotational diffusion anisotropy of proteins from simultaneous analysis of ^{15}N and ^{13}C nuclear spin relaxation. J. Biomolec. NMR 1997, 9, 287–98.
47. Ghose, R., Fushman, D., Cowburn, D. Determination of rotational diffusion tensor of macromolecules in solution from NMR relaxation data with a combination of exact and approximate methods. Application to the determination of interdomain orientation in multidomain proteins. J. Magn. Reson. 2001, 149, 204–17.
48. Kay, L.E., Torchia, D.A., Bax, A. Backbone dynamics of proteins as studied by nitrogen-15 inverse detected heteronuclear NMR spectroscopy: Application to staphylococcal nuclease. Biochemistry 1989, 28, 8972–9.
49. Box, G.E.P., Jenkins, G.M. Time Series Analysis: Forecasting and Control. New York: Holden-Day; 1976.
50. Brockwell, P.J., Davis, R.A. Time Series: Theory and Methods. New York: Springer-Verlag; 1991.
51. Hall, J.B., Fushman, D. Characterization of the overall and local dynamics of a protein with intermediate rotational anisotropy: Differentiating between conformational exchange and anisotropic diffusion in the B3 domain of protein G. J. Biomol. NMR 2003, 27, 261–75.
52. García de la Torre, J., Huertas, M.L., Carrasco, B. Calculation of hydrodynamic properties of globular proteins from their atomic-level structure. Biophys. J. 2000, 78, 719–30.

Applications and Advances of QM/MM Methods in Computational Enzymology

Alessio Lodola*, Christopher J. Woods**, and
Adrian J. Mulholland**,1

Contents

1. INTRODUCTION

Understanding enzyme catalysis is one of the grand challenges of both experimental and computational biochemistry [1]. The past decade has witnessed a dramatic

* Dipartimento Farmaceutico, Università degli Studi di Parma, viale G.P. Usberti 27/A Campus Universitario, I-43100 Parma, Italy

** Centre for Computational Chemistry, School of Chemistry, University of Bristol, Bristol, BS8 1TS, UK
   E-mail: Adrian.Mulholland@bristol.ac.uk (A.J. Mulholland)

1 Corresponding author.

Annual Reports in Computational Chemistry, Vol. 4
ISSN 1574-1400, DOI: 10.1016/S1574-1400(08)00009-1

growth in the availability of high-resolution enzyme structures that have been deposited in the protein data bank (PDB). Computational modeling can provide detailed information about enzyme catalysis that is not readily available from experiment. Given a theoretical model that accurately describes the relative energies along a reaction path, and a method for finding reaction pathways, it is possible to find all of the intermediates and transition states that are relevant for a catalytic process. Once these stationary points have been characterized, simulations can be used to elucidate the details of reaction kinetics. One popular class of computational enzymology methods are those that use a combination of quantum mechanics (QM) and molecular mechanics (MM) to represent the reacting system. In the first part of this paper we review recent applications of QM/MM methods to investigate reactions in a range of enzyme systems. In the second part we review the development of novel QM/MM sampling methods that should help to analyze the effect of conformational dynamics on enzyme activity and catalysis.

2. RECENT APPLICATIONS OF QM/MM METHODS TO ENZYMES

Several methods now exist to model an enzyme-catalyzed reaction. With the expansion of density functional theory (DFT) [2], quantum mechanics (QM) is now capable of describing reactive chemistry for systems involving dozens of atoms with a very respectable level of accuracy [3,4]. However, electronic structure calculations on enzymes (composed of thousands of atoms) require large computational resources, significantly limiting the size of the system that can be treated. On the other hand, molecular mechanics (MM) force fields, such as CHARMM22 [5] or AMBER (PARM99) [6], have been developed to provide a remarkably good description of conformational energetics and non-bonded interactions in large systems [7]. MM methods are generally not applicable to the modeling of chemical reactions, because they are designed and parameterized for chemically stable states [8]. In this context, mixed quantum mechanics/molecular mechanics (QM/MM) approaches have been developed [9] that can readily join QM and MM representations of different parts of a complex condensed-phase system [10]. The combination of these approaches contains the fundamentals necessary to properly describe the potential energy surfaces relevant to enzymatic chemistry, at least to a first approximation [11,12]. In the QM/MM framework, the reactive region of the enzyme active site can be described by a QM method (e.g. semi-empirical, DFT, or ab initio theory) employing a QM region of sufficient size to include the reacting groups [13]. Semi-empirical molecular-orbital techniques such as AM1 and PM3 have the advantage that they can model large systems (hundreds of atoms), but they are often inaccurate and in some cases not easy to use (e.g. for many transition metals) [14]. Reliable calculations have increasingly been made possible by the development of methods based on DFT and on electron correlation approaches, such as the MP2 perturbation method and coupled-cluster theory [15,16].

With the QM/MM model of the system constructed, a straightforward means of modeling approximate reaction paths is the 'adiabatic mapping' or 'coordinate driving' approach. The energy of the system is calculated by minimizing the energy at a series of fixed or restrained values of a reaction coordinate, e.g. the distance between two atoms. This approach has been applied with success to several systems [17–20]. However it is only valid if one conformation of the protein can represent the system during the whole reaction. If only one conformation of the protein appears to be involved in the reaction, a single minimum energy structure of this conformation may adequately represent the many closely related structures making up this conformational state. Minimizing the QM/MM potential energy of such a representative conformation along the reaction coordinate should therefore provide a reasonable approximation of the enthalpic component of the potential of mean force (the free energy profile) for the reaction.

Enzyme catalyzed reactions can also be modeled using the empirical valence bond (EVB) method [21]. EVB methods are central in modeling enzyme reactions and catalysis. They are reviewed well elsewhere [22,12] and so are not the focus of this review.

2.1 Chorismate mutase

Chorismate mutase (CM) catalyzes the Claisen rearrangement of chorismate to prephenate in the shikimic acid pathway used in the biosynthesis of aromatic amino acids. It represents a reference enzyme to explore the fundamentals of catalysis and has been the subject of extensive experimental and computational research. These have shown both that catalysis proceeds without covalent binding of the substrate to the enzyme, and that the uncatalyzed reaction in water proceeds by the same mechanism. This makes CM a particularly convenient target for QM/MM studies.

To date, the majority of applications of QM/MM methods to enzyme reaction modeling have been carried out by using semi-empirical Hamiltonians, which do not predict barrier heights with great accuracy [23]. Recently, higher-level QM/MM calculations have been performed, e.g. at the DFT/MM level [24–26]. The reaction of chorismate to prephenate in CM has also been studied at very high-level, using coupled cluster theory computations that account properly for electron correlation [15]. Coupled cluster theory calculations, namely LCCSD(T0) (the L in the acronym indicates that local approximations were used, and T0 is an approximate triples correction) [27], on CM overcomes the shortcoming of lower level methods. With the unprecedented combination of LCCSD(T0) results and thorough sampling (the structures were optimized at the B3LYP/MM level), the activation enthalpy was calculated with high accuracy. The final value obtained for CM (average 13.3 kcal mol^{-1}, with a root mean square variation of 1.1 kcal mol^{-1} across 16 pathways) can be compared with the experimental value of 12.7 kcal mol^{-1}. Both DFT and standard (particularly) Hartree–Fock (HF) methods failed to reproduce the experimental value, suggesting that a proper treatment of electron correlation is required for quantitative predictions of barrier heights. Similar findings were also reported for another prototypical enzyme, para-hydroxybenzoate hydroxylase (PHBH) [15,28].

2.2 Class A β-lactamases

QM/MM adiabatic mapping has been applied, for example, to study the mechanism of class A β-lactamases, which are an important class of bacterial enzyme involved in antibiotic breakdown. Both the acylation [19,20] and deacylation reactions [29] of the *Escherichia coli* TEM1 β-lactamase with benzylpenicillin have been investigated, and a review [30] of QM/MM applications to the study of β-lacatamases and ureases has been published recently. Structures and interactions with the protein were modeled by a AM1/CHARMM22 QM/MM approach, and calculations of reaction energies were performed at a higher level (B3LYP/6–31+G(d)) to correct the deficiencies of the AM1 semi-empirical method. This approach identified Glu166 as the general base, which acts by deprotonating a structurally conserved water molecule, which in turn activates Ser70 for nucleophilic attack on the β-lactam ring. Protonation of the nitrogen of the antibiotic, performed by Ser130 in combination with Lys73, leads to the formation of the acylenzyme. This mechanism is consistent with experimental kinetic and structural data and the calculated energy barrier (9 kcal mol^{-1}) compares well with the experimental value of 12 kcal mol^{-1}. The first step of the deacylation reaction was also simulated at the B3LYP/6-31G+(d)//AM1-CHARMM22 QM/MM level. These calculations showed that Glu166 acts as a base, deprotonating a conserved water molecule which then acts as a nucleophile, attacking the carbonyl carbon of the acylated serine. The barrier for deacylation was calculated to be 8.7 kcal mol^{-1}, which is very close to the value found for acylation. This indicates that the different steps in antibiotic breakdown have comparable barriers, with no step being clearly rate-limiting. The catalytic rate constant therefore depends on several reaction processes with similar barriers, instead of one dominant rate-determining step. Modeling of the reaction also identified several interactions at the active site, both in the acylation and deacylation processes, which may help in the development of stable β-lactam antibiotics and in designing new lactamase inhibitors.

2.3 Cysteine proteases

Adiabatic mapping has also been applied to investigate cysteine proteases. These enzymes are involved in many diseases, and so represent promising drug targets. Epoxides and aziridines are classical prototypes of cysteine alkylating agents that irreversibly block cysteine proteases by a ring-opening reaction involving an attack of the thiolate group of the cysteine on the three-membered rings. The mechanism of action of these inhibitors has been elucidated by modeling their reaction with cathepsin B at the B3LYP/TZVP//BLYP/TZVP-CHARMM22 level [31]. The calculations suggested that ring opening requires an efficient protonation of the leaving group. Contrary to what was originally proposed, the calculations argued against a direct proton shift from the active site histidine to the inhibitor, but showed that one water molecule is sufficient to establish a very efficient relay system. This relay system allows an easy proton transfer from the active site histidine residue to the inhibitor and it may thus be essential for the activity of both types of inhibitors. These findings may help the design of new analogs based on aziridine and epoxide templates.

2.4 Cytochromes P450

Drug metabolism is another important aspect of the development and optimization of useful, pharmaceutically active compounds, and enzymes involved in
these processes are being studied using QM/MM adiabatic mapping methods.
Cytochromes P450 are key enzymes involved in drug metabolism. A particularly
important reaction is the P450-catalyzed hydroxylation of C–H bonds. This reaction is crucial for drug metabolism as it can be responsible for the activation of
pro-drugs, or influence the pharmacokinetics of many pharmaceuticals. QM/MM
methods can provide a uniquely detailed description of the reactions catalyzed
by cytochrome P450 enzyme family. QM/MM studies of bacterial P450cam have
raised controversial issues about reactivity [32–34]. Different P450 isoenzymes display very different substrate specificity and hydroxylation patterns, which could
be the result of orientation or binding effects, or the intrinsic reactivity of different
locations in the substrate [35]. For example, calculations on models of aromatic
hydroxylation in P450 have identified two different possible orientations of the
substrate (side-on and face-on) [36,37] either or both of which may be important
in the reactions of different drugs in different P450s. This insight from calculations
has led to the development of new structure-activity relationships (SARs) that
may help in the prediction of metabolite products for drugs containing aromatic
fragments. QM/MM modeling of human cytochrome P450 enzymes (including
complexes with the widely-used drugs diclofenac and ibuprofen) demonstrates
the potential of QM/MM methods to deal with practical questions of xenobiotic
metabolism, for example in identifying and analyzing determinants of reactivity [38]. At the same time, future developments in the field of pharmacogenetics
will require models to predict the effects of genetic variation on the activity and
specificity of enzymes responsible for drug metabolism.

2.5 Fatty acid amide hydrolase (FAAH)

Other applications of QM/MM methods relevant for medicinal chemistry concern
fatty acid amide hydrolase (FAAH), a key enzyme involved in endocannabinoid
metabolism, and a promising target for the treatment of central and peripheral
nervous system disorders, such as anxiety, pain and depression [39,40]. The mechanism of the first crucial steps of the acylation reaction catalyzed by FAAH, with
the substrate oleamide, has been recently modeled with a PM3/CHARMM22
QM/MM potential, with DFT calculations for a reliable description of the reaction energetics [41]. The calculations revealed a novel mechanism of nucleophile activation showing that the general base Lys142 and the proton shuttle
Ser217 cooperate to activate the key nucleophile Ser241. Characterization of the
B3LYP/6-31G+(d)//PM3-CHARMM22 potential energy surface for the reaction
indicated that the activation of the nucleophile is the rate limiting step, giving
a barrier of 18 kcal mol^{-1}, close to the experimentally deduced activation barrier (16 kcal mol^{-1}). These simulations identified crucial interactions at the active
site, highlighting the role played by the oxyanion hole, which in FAAH is composed by four consecutive residues (Ile238, Gly239, Gly240 and Ser241) arranged

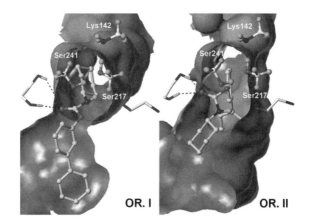

FIGURE 9.1 Binding modes for the inhibitor URB524 in the active site of the enzyme fatty acid amide hydrolase (FAAH), showing both orientations I (left) and II (right) used for QM/MM modeling [43]. Modeling the mechanism of reaction of this covalent inhibitor identified orientation II as the productive binding mode, with the carbamic group of URB524 placed close to the nucleophilic hydroxyl group of Ser241.

in a hairpin-like loop. Comparison between QM/MM and *in vacuo* QM calculations showed that the highest level of stabilization along the pathway occurs at the tetrahedral intermediate, due to the hydrogen bonds between the negatively charged oxygen of oleamide and NH groups of the oxyanion hole residues (i.e. the tetrahedral intermediate is the species most stabilized by the enzyme during the reaction). The active site is well organized to achieve this stabilization, and so to catalyze the reaction effectively.

The interaction between FAAH and carbamic acid aryl ester inhibitors has also been investigated using QM/MM methods. This class of compounds, including the compound URB524, has been shown to inactivate FAAH by carbamoylation of the active nucleophile Ser241 [42]. In general, structure-based drug design depends on the accuracy of ligand docking, and the ability to identify relevant binding modes. URB524 and its derivatives can be docked within the FAAH catalytic site in two possible orientations (called orientation I and II), both placing the carbamic group close to Ser241 (Figure 9.1). Traditional computational tools employed in drug discovery, such as docking and scoring (e.g. by classical interaction energies), failed to clearly discriminate between these two binding orientations [44,45].

The QM/MM approach was applied to model the formation of the covalent inhibitor complex, employing the PM3/CHARMM22 potential with B3LYP/6-31G+(d) energy corrections for a reliable description of the reaction energetics. Potential energy surfaces were calculated for each binding orientation and transition states and intermediates were identified along the reaction profiles. The calculations clearly showed that the carbamoylation in orientation II is energetically preferred, thus identifying II as the productive binding mode [43]. Similar pathways were obtained using other snapshots taken from molecular dynamics simulations, suggesting that the preference observed for binding orientation II is not affected by the starting conformation. These results provide a theoretical basis for SAR interpretation of URB524 analogs, highlighting the useful contribution that mechanistic modeling of enzymes can give in identifying productive binding modes for covalent inhibitors. QM/MM modeling of reaction mechanisms for covalent inhibitors in enzyme targets has the potential to provide detailed information for drug design in cases where traditional docking alone may fail.

2.6 Recent applications of QM/MM to other enzyme systems

Limited space means that it is not possible to review all of the many recent applications of QM/MM methods in computational enzymology. QM/MM is now a mainstream methodology and is the subject of several excellent reviews [46,12,47, 48]. Other excellent examples of its application that are not covered in this review include the investigation of proton transfer in carbonic anhydrase [49], the hydrolysis of paraoxon by phosphotriesterase [50], investigation of the catalytic pathway of cathepsin K [51] (also a cysteine protease, like cathepsin B described above), investigations of the catalytic pathway of catechol-O-methyltransferase [52] and the modeling of hydride transfer in xylose isomerase [53]. QM/MM methods have also been used to study protein-ligand interactions, e.g. investigating the protein-ligand interactions of a range of HIV-1 integrase inhibitors [54].

3. ENZYME CONFORMATIONAL CHANGES AND FLUCTUATIONS: EFFICIENT SAMPLING IN QM/MM SIMULATIONS

There is growing evidence that an account must in many cases be made for the effects of protein conformational change when performing computational enzymology calculations. Conformational change on a small scale (e.g. of a small number of amino acid side chains) may occur during catalysis. A recent investigation on the role of conformational fluctuations on catalysis has been reported for the first step of the acylation reaction between FAAH and oleamide. Potential energy surfaces (PESs) were calculated at the B3LYP/6-31+G(d)//PM3-CHARMM22 level for multiple conformations of the enzyme-substrate complex. The results showed that geometrical fluctuations of the active site can significantly affect the overall energetic barrier. Although conformational fluctuations did not affect the general shape of the PESs, consistency between experimental and calculated barriers is observed only with a specific arrangement of the reactants [55]. These findings strongly suggest that the employment of different protein conformations can be essential for a meaningful determination of the energetic of enzymatic reactions. The effect of small conformational changes was also observed by Warshel and co-workers [56], who investigated the conformational dependence of activation energies calculated using adiabatic mapping. They investigated a hypothetical reaction by running adiabatic mapping calculations on several enzyme conformations taken from a molecular dynamics trajectory. Large differences were observed for both the energy of the ground state (which varied by up to 30 kcal mol^{-1})

and for the activation energy (which varied by up to 6 kcal mol^{-1}). These results demonstrate clearly that small conformational changes can have a significant influence on the calculated activation energy, and that extensive conformational sampling may be required to average out their effects.

It has also been observed that large-scale protein conformational change can be the rate-determining step of enzyme catalytic turnover. For example, Wolf-Watz et al. performed an NMR study of mesophilic and thermophilic homologs of adenylate kinase [57] to study the role of protein dynamics in enzymatic turnover. They demonstrated that the rate-limiting step was a protein conformational change (the opening of the nucleotide binding lids), and that the reduced activity of the hyperthermophilic homolog at ambient temperatures was caused solely by a slower lid-opening rate [57]. This suggests that the enzyme has evolved to the point where the chemical reaction is very fast, and the limitation now to the enzyme's catalytic power is its ability to change conformation [57]. The importance of large scale conformational change is now also being shown computationally. Pentikäinen et al. [58] have recently reported simulations that suggest that complex protein dynamics are central to the function of human scavenger decapping enzyme (DcpS). The standard view has been that substrate binding caused the dimer to change from a closed symmetric conformation to an asymmetric open conformation. The simulations suggested, however, that the apo-form of the enzyme undergoes a continuous process of conformational change, with one side opening and the other closing in a clearly cooperative fashion [58]. This conformational change occurred over a timescale of approximately 4–13 ns. This observation may be of wide relevance, as many enzymes function as dimers, and may undergo similar conformational changes in solution, something that X-ray crystal structures alone cannot show directly [58].

3.1 Direct sampling methods

The effects of protein conformational change can be included via conformational sampling of the QM/MM Hamiltonian. One well-established method is umbrella sampling [59], which works by combining the results of several molecular dynamics simulations that are restrained to sample overlapping regions across the reaction coordinate. For example, Bowman et al. [60] used umbrella sampling to investigate M1-1 glutathione S-transferase, an enzyme that plays an important role in the detoxification of a large variety of xenobiotic compounds in mammals. Activation free energies calculated by Bowman et al. for this enzyme (using a specifically parameterized AM1-SRP/CHARMM22 method in umbrella sampling molecular dynamics simulations) agreed well with experiment. While umbrella sampling was, in this case, successful, its main drawback is that the amount of conformational sampling is limited by the significant computational expense of the evaluation of the QM forces. This has limited QM/MM umbrella sampling applications to trajectory lengths that are typically measured in picoseconds (30 ps at each value of the reaction coordinate, in the case of Bowman et al.), generally using only semi-empirical QM Hamiltonians (e.g. AM1 or PM3).

Jorgensen and co-workers have developed an efficient method of sampling a QM/MM Hamiltonian that enables them to increase the quality of conformational sampling [61–63]. They use the Monte Carlo (MC) [64] method to explore conformational space. MC works by performing small moves of randomly chosen molecules (or parts of molecules). Millions of such moves are required to accurately converge free energy averages. This presents a problem if a standard electronic embedding [9,12] is used to calculate the electrostatic interaction between the QM and MM atoms, as every MC move would require a QM energy calculation. Jorgensen and co-workers solve this problem by calculating the electrostatic QM/MM interaction using partial charges on the QM atoms that are updated dynamically throughout the simulation. They have developed a method [61] to obtain atomic partial charges efficiently from a QM calculation that are compatible with the partial charges from the standard OPLS [65,66] all-atom forcefield. The charges are calculated using the charge model 1 (CM1A) [67] analysis of an AM1 semi-empirical QM calculation. Jorgensen and co-workers have successfully used this method to study solution-phase Diels–Alder reactions [68], and to study the enzyme-catalyzed Claisen rearrangement reaction of chorismate to prephenate in CM [62]. Jorgensen and co-workers have since adapted this method [69–72] to use the PDDG/PM3 semi-empirical QM Hamiltonian [73], using the CM3 [74] method to extract charges. They have also used this method to investigate oleamide and methyl oleate hydrolysis catalyzed by FAAH [39], and obtained results that compared well with the barriers predicted in the adiabatic mapping study [41,43] discussed in the last section.

3.2 Indirect sampling methods

The methods described above include the effects of conformational change via explicit sampling of the QM/MM Hamiltonian. The cost of evaluating the QM energy makes these methods computationally demanding. An alternative to sampling the QM/MM Hamiltonian directly is to perform the sampling using an MM Hamiltonian, and to then perform QM/MM calculations using the resulting MM ensemble. Warshel and co-workers have, over a series of pioneering papers [75,76], developed a method that allows an MM ensemble to be used to calculate, in theory, exact QM/MM free energies. The aim of this method is to calculate the free energy difference between two systems, A and B. For example, system A could be a substrate bound to an enzyme, while system B could be the transition state. The free energy difference between these two corresponds to the activation free energy of the enzyme catalyzed reaction. Warshel and co-workers calculated the relative free energy of A and B by first using a MM type potential. Because an MM potential was used, molecular dynamics sampling was efficient, and therefore a large ensemble, and well-converged relative free energy were calculated. This relative free energy, $\Delta G_{MM}(A \rightarrow B)$, can only be as good as the MM potential used during the calculation. Warshel and co-workers solve this problem by then using the MM ensembles to calculate the difference in free energy between the QM and MM representations of A and B. In essence, Warshel and co-workers calculated the free energy error associated with using the MM forcefield. By calculating these errors, Warshel and co-workers were able to correct $\Delta G_{MM}(A \rightarrow B)$ so that it was formally equal to $\Delta G_{QM}(A \rightarrow B)$ [76,75] (see Figure 9.2). The correction free energies were

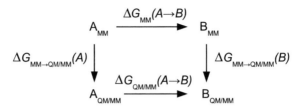

FIGURE 9.2   The free energy cycle [75,76] used to calculate the QM/MM free energy difference between systems A and B, $\Delta G_{QM/MM}(A \to B)$. The free energy difference between A and B is first estimated using an approximate potential (e.g. an MM potential), giving $\Delta G_{MM}(A \to B)$. This is then corrected to the QM/MM value by calculating the free energy necessary to perturb system A from MM to QM/MM ($\Delta G_{MM \to QM/MM}(A)$) and the free energy to perturb system B from MM to QM/MM ($\Delta G_{MM \to QM/MM}(B)$).

calculated by generating ensembles for systems A and B using the MM model. The difference in energy between the QM and MM models was calculated for a subset of each ensemble, and the difference between these energies used as input to a single-step free energy perturbation (FEP) [77,78] between the MM model (the FEP reference state) and the QM model (the FEP perturbed state). As long as the MM model is a good approximation of the QM model, i.e. the phase space overlap of the two models is good, then the average calculated via the FEP equation will converge to an accurate estimate of the correction free energy. The key advantage of this method is that all of the thermodynamic sampling is performed using only the MM model of the system. QM or QM/MM calculations are run in parallel with the MM sampling to estimate the correction free energies. This QM/MM method has been used to study a variety of systems [79–82], including chorismate mutase [83].

Schemes for efficient QM/MM sampling developed for solvation or binding free energy calculations also hold great promise for computational enzymology. For example, we have recently presented a QM/MM free energy method [84] that uses efficient Monte Carlo sampling and Hamiltonian replica exchange [85,86] algorithms to calculate the correction free energies required by the Warshel free energy cycle. The method was tested by calculating the relative hydration free energy of water and methane at the MP2/AVDZ/OPLS level. Extension of this method to the calculation of activation free energies is promising.

4. CONCLUSION

There has been significant recent development in the application of QM/MM methods for computational enzymology, whereby both the detail of the physical model of the enzyme system, and the quality of conformational sampling have improved dramatically. This has now led to computational simulations of certain enzyme systems that are sufficiently accurate to allow direct comparison with experiment. Computer simulations have now reached a level of accuracy, for some systems, that permits their use as a means of verifying or dismissing proposed enzyme reaction mechanisms. This is a significant development, with potential benefits both for the field of enzymology, and for rational drug design.

ACKNOWLEDGMENTS

A.L. is grateful to Prof. Marco Mor (Università degli Studi di Parma) for fruitful discussions. C.J.W. and A.J.M. are grateful to the EPSRC for funding.

REFERENCES

1. Garcia-Viloca, M., Gao, J., Karplus, M., Truhlar, D.G. How enzymes work: Analysis by modern rate theory and computer simulations. Science 2004, 303(5655), 186–95.
2. Koch, W., Holthausen, M.C. A Chemist's Guide to Density Functional Theory. Verlag GmbH: Wiley-VCH; 2001.
3. Himo, F., Siegbahn, P.E.M. Quantum chemical studies of radical-containing enzymes. Chem. Rev. 2003, 103(6), 2421–56.
4. Harvey, J.N., Aggarwal, V.K., Bathelt, C.M., Carreon-Macedo, J.L., Gallagher, T., Holzmann, N., Mulholland, A.J., Robiette, R. QM and QM/MM studies of selectivity in organic and bio-organic chemistry. J. Phys. Org. Chem. 2006, 19(8–9), 608–15.
5. MacKerell, A.D., Bashford, D., Bellott, M., Dunbrack, R.L., Evanseck, J.D., Field, M.J., Fischer, S., Gao, J., Guo, H., Ha, S., Joseph-McCarthy, D., Kuchnir, L., Kuczera, K., Lau, F.T.K., Mattos, C., Michnick, S., Ngo, T., Nguyen, D.T., Prodhom, B., Reiher, W.E., Roux, B., Schlenkrich, M., Smith, J.C., Stote, R., Straub, J., Watanabe, M., Wiorkiewicz-Kuczera, J., Yin, D., Karplus, M. All-atom empirical potential for molecular modeling and dynamics studies of proteins. J. Phys. Chem. B 1998, 102(18), 3586–616.
6. Wang, J.M., Cieplak, P., Kollman, P.A. How well does a restrained electrostatic potential (RESP) model perform in calculating conformational energies of organic and biological molecules? J. Comput. Chem. 2000, 21(12), 1049–74.
7. Adcock, S.A., McCammon, J.A. Molecular dynamics: Survey of methods for simulating the activity of proteins. Chem. Rev. 2006, 106(5), 1589–615.
8. Karplus, M., Kuriyan, J. Molecular dynamics and protein function. Proc. Natl. Acad. Sci. USA 2005, 102(19), 6679–85.
9. Warshel, A., Levitt, M. Theoretical studies of enzymic reactions—Dielectric, electrostatic and steric stabilization of carbonium-ion in reaction of lysozyme. J. Mol. Biol. 1976, 103(2), 227–49.
10. Field, M.J., Bash, P.A., Karplus, M. A combined quantum-mechanical and molecular mechanical potential for molecular-dynamics simulations. J. Comput. Chem. 1990, 11(6), 700–33.
11. Mulholland, A.J. Computational enzymology: Modelling the mechanisms of biological catalysts. Biochem. Soc. Trans. 2008, 36, 22–6.
12. Warshel, A. Computer simulations of enzyme catalysis: Methods, progress, and insights. Annu. Rev. Biophys. Biomolec. Struct. 2003, 32, 425–43.
13. Friesner, R.A., Guallar, V. Ab initio quantum chemical and mixed quantum mechanics/molecular mechanics (QM/MM) methods for studying enzymatic catalysis. Annu. Rev. Phys. Chem. 2005, 56, 389–427.
14. Ridder, L., Mulholland, A.J. Modeling biotransformation reactions by combined quantum mechanical/molecular mechanical approaches: From structure to activity. Curr. Top. Med. Chem. 2003, 3(11), 1241–56.
15. Claeyssens, F., Harvey, J.N., Manby, F.R., Mata, R.A., Mulholland, A.J., Ranaghan, K.E., Schutz, M., Thiel, S., Thiel, W., Werner, H.J. High-accuracy computation of reaction barriers in enzymes. Angew. Chem. Int. Edit. 2006, 45(41), 6856–9.
16. Mulholland, A.J. Chemical accuracy in QM/MM calculations on enzyme-catalysed reactions. Chem. Central. J. 2007, 1, 19.

17. Mulholland, A.J. The QM/MM approach to enzymatic reactions. In: Theoretical Biochemistry. Elsevier; 2001, pp. 597–653.
18. Mulholland, A.J. Modelling enzyme reaction mechanisms, specificity and catalysis. Drug Discov. Today 2005, 10(20), 1393–402.
19. Hermann, J.C., Ridder, L., Mulholland, A.J., Holtje, H.D. Identification of GLU166 as the general base in the acylation reaction of class A β-lactamases through QM/MM modeling. J. Am. Chem. Soc. 2003, 125(32), 9590–1.
20. Hermann, J.C., Hensen, C., Ridder, L., Mulholland, A.J., Holtje, H.D. Mechanisms of antibiotic resistance: QM/MMM modeling of the acylation reaction of a class A β-lactamase with benzylpenicillin. J. Am. Chem. Soc. 2005, 127(12), 4454–65.
21. Åqvist, J., Warshel, A. Simulation of enzyme-reactions using valence-bond force-fields and other hybrid quantum-classical approaches. Chem. Rev. 1993, 93(7), 2523–44.
22. Villa, J., Bentzien, J., Gonzalez-Lafont, A., Lluch, J.M., Bertran, J., Warshel, A. Effective way of modeling chemical catalysis: Empirical valence bond picture of role of solvent and catalyst in alkylation reactions. J. Comput. Chem. 2000, 21(8), 607–25.
23. Lyne, P.D., Mulholland, A.J., Richards, W.G. Insights into chorismate mutase catalysis from a combined QM/MM simulation of the enzyme reaction. J. Am. Chem. Soc. 1995, 117(45), 11345–50.
24. Claeyssens, F., Ranaghan, K.E., Manby, F.R., Harvey, J.N., Mulholland, A.J. Multiple high-level QM/MM reaction paths demonstrate transition-state stabilization in chorismate mutase: Correlation of barrier height with transition-state stabilization. Chem. Commun. 2005(40), 5068–70.
25. Crespo, A., Marti, M.A., Estrin, D.A., Roitberg, A.E. Multiple-steering QM/MM calculation of the free energy profile in chorismate mutase. J. Am. Chem. Soc. 2005, 127(19), 6940–1.
26. Marti, S., Moliner, V., Tunon, M., Williams, I.H. Computing kinetic isotope effects for chorismate mutase with high accuracy. A new DFT/MM strategy. J. Phys. Chem. B 2005, 109(9), 3707–10.
27. Schütz, M. Low-order scaling local electron correlation methods. III. Linear scaling local perturbative triples correction (T). J. Chem. Phys. 2000, 113(22), 9986–10001.
28. Mata, R.A., Werner, H.J., Thiel, S., Thiel, W. Toward accurate barriers for enzymatic reactions: QM/MM case study on p-hydroxybenzoate hydroxylase. J. Chem. Phys. 2008, 128(2), 251–4.
29. Hermann, J.C., Ridder, L., Hotje, H.D., Mulholland, A.J. Molecular mechanisms of antibiotic resistance: QM/MM modelling of deacylation in a class A β-lactamase. Org. Biomol. Chem. 2006, 4(2), 206–10.
30. Estiu, G., Suarez, D., Merz, K.M. Quantum mechanical and molecular dynamics simulations of ureases and Zn β-lactamases. J. Comput. Chem. 2006, 27(12), 1240–62.
31. Mladenovic, M., Schirmeister, T., Thiel, S., Thiel, W., Engels, B. The importance of the active site histidine for the activity of epoxide- or aziridine-based inhibitors of cysteine proteases. Chem. Med. Chem. 2007, 2(1), 120–8.
32. Schoneboom, J.C., Cohen, S., Lin, H., Shaik, S., Thiel, W. Quantum mechanical/molecular mechanical investigation of the mechanism of C–H hydroxylation of camphor by cytochrome p450(CAM): Theory supports a two-state rebound mechanism. J. Am. Chem. Soc. 2004, 126(12), 4017–34.
33. Guallar, V., Baik, M.H., Lippard, S.J., Friesner, R.A. Peripheral heme substituents control the hydrogen-atom abstraction chemistry in cytochromes p450. Proc. Natl. Acad. Sci. USA 2003, 100(12), 6998–7002.
34. Zurek, J., Foloppe, N., Harvey, J.N., Mulholland, A.J. Mechanisms of reaction in cytochrome p450: Hydroxylation of camphor in p450CAM. Org. Biomol. Chem. 2006, 4(21), 3931–7.
35. de Groot, M.J., Kirton, S.B., Sutcliffe, M.J. In silico methods for predicting ligand binding determinants of cytochromes p450. Curr. Top. Med. Chem. 2004, 4(16), 1803–24.
36. Bathelt, C.M., Ridder, L., Mulholland, A.J., Harvey, J.N. Aromatic hydroxylation by cytochrome p450: Model calculations of mechanism and substituent effects. J. Am. Chem. Soc. 2003, 125(49), 15004–5.
37. Bathelt, C.M., Ridder, L., Mulholland, A.J., Harvey, J.N. Mechanism and structure–reactivity relationships for aromatic hydroxylation by cytochrome p450. Org. Biomol. Chem. 2004, 2(20), 2998–3005.
38. Bathelt, C.M., Zurek, J., Mulholland, A.J., Harvey, J.N. Electronic structure of compound I in human isoforms of cytochrome p450 from QM/MM modeling. J. Am. Chem. Soc. 2005, 127(37), 12900–8.

39. Tubert-Brohman, I., Acevedo, O., Jorgensen, W.L. Elucidation of hydrolysis mechanisms for fatty acid amide hydrolase and its LYS142ALA variant via QM/MM simulations. J. Am. Chem. Soc. 2006, 128(51), 16904–13.
40. Piomelli, D., Tarzia, G., Duranti, A., Tontini, A., Mor, M., Compton, T.R., Dasse, O., Monaghan, E.P., Parrott, J.A., Putman, D. Pharmacological profile of the selective FAAH inhibitor KDS-4103 (URB597). CNS Drug Rev. 2006, 12(1), 21–38.
41. Lodola, A., Mor, M., Hermann, J.C., Tarzia, G., Piomelli, D., Mulholland, A.J. QM/MM modelling of oleamide hydrolysis in fatty acid amide hydrolase (FAAH) reveals a new mechanism of nucleophile activation. Chem. Commun. 2005(35), 4399–401.
42. Alexander, J.P., Cravatt, B.F. Mechanism of carbamate inactivation of FAAH: Implications for the design of covalent inhibitors and in vivo functional probes for enzymes. Chem. Biol. 2005, 12(11), 1179–87.
43. Lodola, A., Mor, M., Rivara, S., Christov, C., Tarzia, G., Piomelli, D., Mulholland, A.J. Identification of productive inhibitor binding orientation in fatty acid amide hydrolase (FAAH) by QM/MM mechanistic modelling. Chem. Commun. 2008(2), 214–6.
44. Mor, M., Rivara, S., Lodola, A., Plazzi, P.V., Tarzia, G., Duranti, A., Tontini, A., Piersanti, G., Kathuria, S., Piomelli, D. Cyclohexylcarbamic acid 3'- or 4'-substituted biphenyl-3-yl esters as fatty acid amide hydrolase inhibitors: Synthesis, quantitative structure–activity relationships, and molecular modeling studies. J. Med. Chem. 2004, 47(21), 4998–5008.
45. Tarzia, G., Duranti, A., Gatti, G., Piersanti, G., Tontini, A., Rivara, S., Lodola, A., Plazzi, P.V., Mor, M., Kathuria, S., Piomelli, D. Synthesis and structure–activity relationships of FAAH inhibitors: Cyclohexylcarbamic acid biphenyl esters with chemical modulation at the proximal phenyl ring. Chem. Med. Chem. 2006, 1(1), 130–9.
46. Gao, J.L., Truhlar, D.G. Quantum mechanical methods for enzyme kinetics. Annu. Rev. Phys. Chem. 2002, 53, 467–505.
47. Riccardi, D., Schaefer, P., Yang, Y., Yu, H.B., Ghosh, N., Prat-Resina, X., Konig, P., Li, G.H., Xu, D.G., Guo, H., Elstner, M., Cui, Q. Development of effective quantum mechanical/molecular mechanical (QM/MM) methods for complex biological processes. J. Phys. Chem. B 2006, 110(13), 6458–69.
48. Lin, H., Truhlar, D.G. QM/MM: What have we learned, where are we, and where do we go from here?. Theor. Chem. Acc. 2007, 117(2), 185–99.
49. Riccardi, D., Konig, P., Guo, H., Cui, Q. Proton transfer in carbonic anhydrase is controlled by electrostatics rather than the orientation of the acceptor. Biochemistry 2008, 47(8), 2369–78.
50. Wong, K.Y., Gao, J. The reaction mechanism of paraoxon hydrolysis by phosphotriesterase from combined QM/MM simulations. Biochemistry 2007, 46(46), 13352–69.
51. Ma, S., Devi-Kesavan, L.S., Gao, J. Molecular dynamics simulations of the catalytic pathway of a cysteine protease: A combined QM/MM study of human cathepsin K. J. Am. Chem. Soc. 2007, 129(44), 13633–45.
52. Roca, M., Moliner, V., Ruiz-Pernia, J.J., Silla, E., Tunon, I. Activation free energy of catechol O-methyltransferase corrections to the potential of mean force. J. Phys. Chem. A 2006, 110(2), 503–9.
53. Garcia-Viloca, M., Alhambra, C., Truhlar, D.G., Gao, J.L. Hydride transfer catalyzed by xylose isomerase: Mechanism and quantum effects. J. Comput. Chem. 2003, 24(2), 177–90.
54. Alves, C.N., Marti, S., Castillo, R., Andres, J., Moliner, V., Tunon, I., Silla, E. A quantum mechanics/molecular mechanics study of the protein–ligand interaction for inhibitors of HIV-1 integrase. Chem. Eur. J. 2007, 13(27), 7715–24.
55. Lodola, A., Mor, M., Zurek, J., Tarzia, G., Piomelli, D., Harvey, J.N., Mulholland, A.J. Conformational effects in enzyme catalysis: Reaction via a high energy conformation in fatty acid amide hydrolase. Biophys. J. 2007, 92(2), L20–2.
56. Klahn, M., Braun-Sand, S., Rosta, E., Warshel, A. On possible pitfalls in ab initio quantum mechanics/molecular mechanics minimization approaches for studies of enzymatic reactions. J. Phys. Chem. B 2005, 109(32), 15645–50.
57. Wolf-Watz, M., Thai, V., Henzler-Wildman, K., Hadjipavlou, G., Eisenmesser, E.Z., Kern, D. Linkage between dynamics and catalysis in a thermophilic-mesophilic enzyme pair. Nat. Struct. Mol. Biol. 2004, 11(10), 945–9.
58. Pentikäinen, U., Pentikäinen, O.T., Mulholland, A.J. Cooperative symmetric to asymmetric conformational transition of the apo-form of scavenger decapping enzyme revealed by simulations. Proteins 2008, 70(2), 498–508.

59. Torrie, G.M., Valleau, J.P. Non-physical sampling distributions in Monte Carlo free-energy estimation—Umbrella sampling. J. Comput. Phys. 1977, 23(2), 187–99.
60. Bowman, A.L., Ridder, L., Rietjens, I.M.C.M., Vervoort, J., Mulholland, A.J. Molecular determinants of xenobiotic metabolism: QM/MM simulation of the conversion of 1-chloro-2,4-dinitrobenzene catalyzed by M1-1 glutathione S-transferase. Biochemistry 2007, 46(21), 6353–63.
61. Kaminski, G.A., Jorgensen, W.L. A quantum mechanical and molecular mechanical method based on CM1A charges: Applications to solvent effects on organic equilibria and reactions. J. Phys. Chem. B 1998, 102(10), 1787–96.
62. Guimaraes, C.R.W., Udier-Blagovic, M., Tubert-Brohman, I., Jorgensen, W.L. Effects of Arg90 neutralization on the enzyme-catalyzed rearrangement of chorismate to prephenate. J. Chem. Theory Comput. 2005, 1(4), 617–25.
63. Guimaraes, C.R.W., Udier-Blagovic, M., Jorgensen, W.L. Macrophomate synthase: QM/MM simulations address the Diels–Alder versus Michael–Aldol reaction mechanism. J. Am. Chem. Soc. 2005, 127(10), 3577–88.
64. Metropolis, N., Rosenbluth, A.W., Rosenbluth, M.N., Teller, A.H., Teller, E. Equation of state calculations by fast computing machines. J. Chem. Phys. 1953, 21(6), 1087–92.
65. Jorgensen, W.L., Madura, J.D., Swenson, C.J. Optimized intermolecular potential functions for liquid hydrocarbons. J. Am. Chem. Soc. 1984, 106(22), 6638–46.
66. Jorgensen, W.L., Maxwell, D.S., TiradoRives, J. Development and testing of the OPLS all-atom force field on conformational energetics and properties of organic liquids. J. Am. Chem. Soc. 1996, 118(45), 11225–36.
67. Storer, J.W., Giesen, D.J., Cramer, C.J., Truhlar, D.G. Class-IV charge models—A new semiempirical approach in quantum-chemistry. J. Comput. Aided Mol. Des. 1995, 9(1), 87–110.
68. Chandrasekhar, J., Shariffskul, S., Jorgensen, W.L. QM/MM simulations for Diels–Alder reactions in water: Contribution of enhanced hydrogen bonding at the transition state to the solvent effect. J. Phys. Chem. B 2002, 106(33), 8078–85.
69. Udier-Blagovic, M., De Tirado, P.M., Pearlman, S.A., Jorgensen, W.L. Accuracy of free energies of hydration using CM1 and CM3 atomic charges. J. Comput. Chem. 2004, 25(11), 1322–32.
70. Acevedo, O., Jorgensen, W.L. Cope elimination: Elucidation of solvent effects from QM/MM simulations. J. Am. Chem. Soc. 2006, 128(18), 6141–6.
71. Acevedo, O., Jorgensen, W.L. Medium effects on the decarboxylation of a biotin model in pure and mixed solvents from QM/MM simulations. J. Org. Chem. 2006, 71(13), 4896–902.
72. Acevedo, O., Jorgensen, W.L. Understanding rate accelerations for Diels–Alder reactions in solution using enhanced QM/MM methodology. J. Chem. Theory Comput. 2007, 3(4), 1412–9.
73. Repasky, M.P., Chandrasekhar, J., Jorgensen, W.L. PDDG/PM3 and PDDG/MNDO: Improved semiempirical methods. J. Comput. Chem. 2002, 23(16), 1601–22.
74. Thompson, J.D., Cramer, C.J., Truhlar, D.G. Parameterization of charge model 3 for AM1, PM3, BLYP and B3LYP. J. Comput. Chem. 2003, 24(11), 1291–304.
75. Muller, R.P., Warshel, A. Ab initio calculations of free-energy barriers for chemical-reactions in solution. J. Phys. Chem. 1995, 99(49), 17516–24.
76. Štrajbl, M., Hong, G.Y., Warshel, A. Ab initio QM/MM simulation with proper sampling: "First principle" calculations of the free energy of the autodissociation of water in aqueous solution. J. Phys. Chem. B 2002, 106(51), 13333–43.
77. Zwanzig, R.W. High-temperature equation of state by a perturbation method. 1. Nonpolar gases. J. Chem. Phys. 1954, 22(8), 1420–6.
78. Jorgensen, W.L., Blake, J.F., Buckner, J.K. Free-energy of TIP4P water and the free-energies of hydration of CH4 and Cl-from statistical perturbation theory. Chem. Phys. 1989, 129(2), 193–200.
79. Wood, R.H., Yezdimer, E.M., Sakane, S., Barriocanal, J.A., Doren, D.J. Free energies of solvation with quantum mechanical interaction energies from classical mechanical simulations. J. Chem. Phys. 1999, 110(3), 1329–37.
80. Ming, Y., Lai, G.L., Tong, C.H., Wood, R.H., Doren, D.J. Free energy perturbation study of water dimer dissociation kinetics. J. Chem. Phys. 2004, 121(2), 773–7.
81. Rod, T.H., Ryde, U. Accurate QM/MM free energy calculations of enzyme reactions: Methylation by catechol O-methyltransferase. J. Chem. Theory Comput. 2005, 1(6), 1240–51.
82. Rosta, E., Klahn, M., Warshel, A. Towards accurate ab initio QM/MM calculations of free-energy profiles of enzymatic reactions. J. Phys. Chem. B 2006, 110(6), 2934–41.

83. Štrajbl, M., Shurki, A., Kato, M., Warshel, A. Apparent NAC effect in chorismate mutase reflects electrostatic transition state stabilization. J. Am. Chem. Soc. 2003, 125(34), 10228–37.

84. Woods, C.J., Manby, F.R., Mulholland, A.J. An efficient method for the calculation of QM/MM free energies. J. Chem. Phys. 2008, 128(1), 014109.

85. Fukunishi, H., Watanabe, O., Takada, S. On the Hamiltonian replica exchange method for efficient sampling of biomolecular systems: Application to protein structure prediction. J. Chem. Phys. 2002, 116(20), 9058–67.

86. Woods, C.J., Essex, J.W., King, M.A. The development of replica-exchange-based free-energy methods. J. Phys. Chem. B 2003, 107(49), 13703–10.

Section 4
Physical Modeling

Section Editor: Jeffry D. Madura

Duquesne University
Department of Chemistry and Biochemistry
Center for Computational Sciences
600 Forbes Ave.
Pittsburg, PA 15282
USA

Stochastic Models for Polymerization Reactions Under Nonequilibrium Conditions

Yanping Qin*, Alexander V. Popov*, and
Rigoberto Hernandez*

Contents

* Center for Computational Molecular Science and Technology, School of Chemistry and Biochemistry, Georgia Institute of Technology, Atlanta, GA 30332-0400, USA
  E-mail: hernandez@chemistry.gatech.edu (R. Hernandez)

Annual Reports in Computational Chemistry, Vol. 4
ISSN 1574-1400, DOI: 10.1016/S1574-1400(08)00010-8

1. INTRODUCTION

Under the typical conditions of equilibrium polymerization [1–16], the environmental response is predominantly stationary and the resulting equilibrium polymer-length distribution is well understood [1–3,6,9–11]. Tobolsky and Eisenberg [4] first treated equilibrium polymerization using mechanistic master equations. De Gennes [7] and des Cloiseaux [8] used renormalization group theory in interpreting continuum models of polymerization as a phase transition between small and high polymers. This interpretation was further validated by Wheeler and Pfeuty [12], who showed that Scott's generalization [5] of the Tobolsky and Eisenberg model is equivalent to an Ising spin magnet in the limit that the spin vector dimension goes to zero. Several groups [17–21] have studied equilibrium distributions and phase diagrams of living polymers [22–24]—that is, addition polymerization in which the active sites remain unterminated or active—by exploiting the isomorphism between continuum and lattice models. Living polymerization [25–39] has been a focus of many statistical models because the sequence distribution equilibrates at long times.

In typical polymerization schemes, macroscopic kinetic equations provide a reasonably accurate reaction mechanism [40,41]. But materials, in which the polymerization—either because of vulcanization in cross-linking reactions or because of aggregative assembly in chain growth—leads to substantial changes in the solvation and hence in the reaction rates, exhibit strongly nonlinear (complex) dynamics. For example, nonstationary response has been seen in thermosetting [42–45] polymers. Thermosetting reactions play an important role in reaction-injected molding [46,47] and have been the subject of large-scale finite-element calculations with semi-empirical kinetic and visco-elastic equations [48]. Throughout such nonequilibrium processes, the viscosity is changing, and must therefore provide time-dependent reaction environments (cages). Thus, though the microscopic elongation reaction in the vacuum may be independent of molecular weight, the average environment of the cage—the potential of mean force—for polymerization will differ as the population of the molecular weight distribution shifts towards higher polymers. The diffusion of the condensate products away from the reaction sites as well as the diffusion of the reactants towards each other has been included in the kinetic models only in an averaged sense. However, as the viscosity changes for the reasons explained above, these diffusion processes will also be affected time-dependently.

Thus far, we have described the time-dependent nature of polymerizing environments both through stochastic [49–51] and lattice [52,53] models capable of addressing this kind of dynamics in a complex environment. The current article focuses on the former approach, but now rephrases the earlier justification of the use of the irreversible Langevin equation, iGLE, to the polymerization problem in the context of kinetic models, and specifically the chemical stochastic equation. The nonstationarity in the solvent response due to the collective polymerization of the dense solvent now appears naturally. This leads to a clear recipe for the construction of the requisite terms in the iGLE. Namely the potential of mean force and the friction kernel as described in Section 3. With these tools in hand, the iGLE is used to analyze the kinetics of the polymer size distribution (and other observables) for two distinct living polymer systems in Section 4.

2. POLYMERIZATION KINETICS MODELED BY THE CHEMICAL STOCHASTIC EQUATION

2.1 The CSE for polymerization reactions

The usual kinetic description of polymerization reactions can be written as a series of consecutive reactions,

$$I + M \rightarrow P_1, \tag{1a}$$

$$P_n + M \underset{k_d}{\overset{k_r}{\rightleftharpoons}} P_{n+1}, \tag{1b}$$

in which P_n is an "activated" polymer consisting of n monomers, and the first reaction (1a) denotes the production of the first activated monomer P_1—i.e., the initiation of polymerization. This set of reactions can be trivially generalized by noting that the reaction rates may differ as the polymer changes—i.e., for different n—and as the overall system changes in time. Formally, the recombination and dissociation rates, $k_r = k_r(n, t)$ and $k_d = k_d(n + 1, t)$, are written as functions which depend on the length n ($\geqslant 1$) of the reacting polymer chain and time. This extension, however, is not so trivial in that it potentially affects the kinetics dramatically, and it requires additional theory and/or equations of motion to describe the evolution of the rates.

The kinetic master equations associated with the reaction mechanism of Eq. (1) can be written as

$$\frac{d}{dt}[P_1] = a(t) - k_r(1, t)[M][P_1] + k_d(2, t)[P_2], \tag{2a}$$

$$\frac{d}{dt}[P_2] = k_r(1, t)[M][P_1] - k_d(2, t)[P_2] + k_d(2, t)[P_3] - k_r(2, t)[M][P_2], \tag{2b}$$

$$\dots$$

$$\frac{d}{dt}[P_n] = k_r(n - 1, t)[M][P_{n-1}] - k_d(n, t)[P_n]$$
$$+ k_d(n + 1, t)[P_{n+1}] - k_r(n, t)[M][P_n], \tag{2c}$$

where $[M]$ is the concentration of inactivated monomers and $[P_n]$ is the concentration of the activated polymer chains. The function $a(t)$ represents the rate of activation of monomers by some external (and possibly time-dependent) driving force. Eq. (1a). Its integral multiplied by volume,

$$A = V \int_0^\infty a(t) \, dt,$$

is the total production of activated monomers during the course of reaction. If no such activated monomer is ever deactivated to monomer, it also represents the total number of polymers (of any size) produced in the solution during the course of reaction.

For long polymer chains n can be treated as a continuous variable. The polymer length is connected to n as $R = nL$, where L is the average monomer length. In this case, the expansion of the quantities $k(n \pm 1)[P_{n\pm1}]$ up to second order,

$$k(n \pm 1)[P_{n\pm1}] \approx k(n)[P_n] \pm \frac{\partial}{\partial n}(k(n)[P_n]) + \frac{1}{2}\frac{\partial^2}{\partial^2 n}(k(n)[P_n]), \tag{3}$$

yields (invoking the generalized notation, $P(n,t) \equiv [P_n]$) an auxiliary partial differential equation,

$$\frac{\partial P(n,t)}{\partial t} = \frac{\partial}{\partial n}\left[(k_d(n,t) - k_r(n,t)[M])P(n,t)\right]$$
$$+ \frac{1}{2}\frac{\partial^2}{\partial^2 n}\left[(k_d(n,t) + k_r(n,t)[M])P(n,t)\right] + \delta(n-1)a(t). \tag{4}$$

The initiation mechanism is visible in the last term. The boundary conditions for Eq. (4) read

$$P(\infty,t) = 0, \tag{5a}$$

$$\left.\frac{\partial}{\partial n}P(n,t)\right|_{n=1} = \left.\frac{f(n,t)}{k_BT}P(n,t)\right|_{n=1}, \tag{5b}$$

where the effective force is

$$\frac{f(n,t)}{k_BT} = -2\frac{k_d(n,t) - k_r(n,t)[M]}{k_d(n,t) + k_r(n,t)[M]} - \frac{\partial}{\partial n}\ln(k_d(n,t) + k_r(n,t)[M]). \tag{5c}$$

Note that Eq. (4), written in the form

$$\frac{\partial P(n,t)}{\partial t} = \frac{\partial}{\partial n}\left[\tilde{D}(n,t)\left(\frac{\partial P(n,t)}{\partial n} - \frac{f(n,t)}{k_BT}P(n,t)\right)\right] + \delta(n-1)a(t), \tag{6}$$

describes the diffusional motion along the continuous coordinate n with the force $f(n,t)$ and "diffusion coefficient"

$$\tilde{D}(n,t) = \frac{1}{2}(k_d(n,t) + k_r(n,t)[M]). \tag{7}$$

If we further assume that dynamics of n is a Markov process, then the "diffusional" chemical kinetic equation, Eq. (4) or (6), acquires the form of the chemical stochastic equation [54,55] (CSE),

$$\dot{n} = -(k_d(n,t) - k_r(n,t)[M]) + \sqrt{k_d(n,t) + k_r(n,t)[M]}\xi(t), \tag{8}$$

and should be understood in Itô's sense [56]. The Gaussian stochastic "force," $\xi(t)$, is "δ"-correlated,

$$\langle\xi(t)\xi(t')\rangle = \delta(t-t'), \tag{9}$$

and the initial condition values, t_0 and n_0, are distributed in accordance with the function

$$\rho(n_0, t_0) = \delta(n_0 - 1)\, a(t_0)/A.$$

The requisite conditions for the general applicability of the CSE have been discussed extensively by Gillespie [54,55]. In the present case of a growing polymer chain, this approach is valid in the limit

$$[P_n]V\, dn_r \gg 1 \quad \text{and} \quad [P_n]V\, dn_d \gg 1,$$

where the change in polymer length, dn, during time dt is

$$dn = dn_r - dn_d. \tag{10}$$

The changes dn_r and dn_d correspond to the number n_r of monomers attached to the polymer and the number n_d of dissociated monomers. ($[P_n]V$ is the number of polymers with length n in the solution.) The quantities in the RHS of Eq. (10) are Poisson distributed with averages

$$\langle dn_r \rangle = k_r[M]\, dt \quad \text{and} \quad \langle dn_d \rangle = k_d\, dt. \tag{11}$$

The validity criteria suggests that the time interval dt must be chosen long enough so that many reactions occur during dt for every reaction channel. In the thermodynamic limit, in which the number of corresponding polymer molecules is large, such an approximation is well justified. The Poisson distribution is then approximated by a Gaussian distribution, and, thus,

$$dn_{r,d} = \langle dn_{r,d} \rangle + \sqrt{\langle dn_{r,d} \rangle}\, \xi_{r,d}(t)\sqrt{dt}. \tag{12}$$

The statistically independent Gaussian white-noise processes, ξ_r and ξ_d, are delta-correlated by analogy with Eq. (9). Combining Eqs. (10), (11) and (12), one arrives at Eq. (8).

In a homogeneous stationary environment the distribution of n is close to Poisson, which is a narrow function with the average $\langle n \rangle$ and width $\sqrt{\langle n \rangle}$. Thus, averaging Eq. (8) over the fluctuating force, one gets the usual kinetic equation,

$$\langle \dot{n} \rangle = -k_d(\langle n \rangle, t) + k_r(\langle n \rangle, t)[M]. \tag{13}$$

Below, we suggest two different approaches to extend the CSE formalism, Eq. (8), for the case of nonequilibrium polymerization.

2.2 CSE with stationary rate coefficients

In order to complete the set of equations describing polymerization reactions in dense and concentrated regimes, the rates k_r and k_d must be specified. In the stationary regime, in which the environmental responses to the microscopic motion of the polymer reactants can be assumed to follow the same regression throughout the reaction, these rates are time-independent. The theory of bimolecular reactions in liquids can then be applied at every reaction step.

There exist many different approximations describing various aspects of diffusion-controlled reversible reactions [57–64]. When the concentration of monomers is high enough, the modified rate-equation approach [58] describes polymerization well. Within the framework of this theory, for freely diffusing reagents, the steady-state rate coefficients in liquids take the form

$$k_{d,r} = \frac{\kappa_{d,r}}{1 + \kappa_r/k_D},$$

(14)

where $\kappa_{d(r)}$ is the intrinsic dissociation (association) rate constant and k_D is the diffusional rate coefficient defined through the properties of motion and geometry of the reagents. In the case of two spherically symmetrical reaction centers, A and B, for example, this diffusion rate coefficient is

$$k_D = 4\pi D\sigma,$$

where $D = D_A + D_B$ is their mutual diffusion coefficient and σ is the sum of the particle radii (contact distance).

The critical assumption behind the use of Eq. (14) is that under steady-state conditions, the spacial flows of particles—which control the flows through the reaction channels—are in quasi-equilibrium. When two small particles react, the time for reaching quasi-equilibrium is about the time for the diffusion of one particle around another, σ^2/D. If one of the reaction centers belongs to a polymer molecule, then the slow modes of configurational motion of a molecular chain will not allow them to reach quasi-equilibrium for a long time, thus, leading to time-dependent rate coefficients. This kind of nonstationarity, caused by slow relaxational mechanisms of polymers, leads to the modified reaction rates (14) which become explicitly time-dependent, and will be described below within the framework of the theory of diffusion controlled reactions.

Other nonstationary effects could arise from the changes in the density of the environment that results from the growth of polymer chains. Such a mechanism is connected to the migration properties of monomers and contributes to the rates differently: the reaction rates become time-dependent through the state of the surroundings. In homogeneous systems, the state of environment can be determined by the averaged polymer length, while in inhomogeneous solutions the local environment and, therefore, the length distribution plays an important role. In this paper, we assume that the system is homogeneous, and that all nonstationary responses are homogeneous in space.

2.3 CSE with nonstationary rate coefficients

We now construct the rate expressions in the nonstationary case. The Laplace transform of the modified rates $k_{d,r}(t)$ yields [62]

$$s\tilde{k}_{d,r}(s) = \frac{\kappa_{d,r}}{1 + \kappa_r\tilde{p}(s)},$$

(15)

recalling that the Laplace transform of a function $f(t)$ is $\tilde{f}(s) = \int_0^\infty \exp(-st)f(t)\,dt$. The function $\tilde{p}(s)$ is the Laplace transform of the partial probability $p(t)$ for finding two particles in contact at time t after being initially in contact.

The partial probability density $p(t)$ can be expressed directly as

$$p(t) = v_c^{-1} \int d^3r \psi (r_2' - r_2) \int d^3r' \psi (r_1' - r_1) G_M(r_2', r_1'; t) G_P(r_2, r_1; t) \quad (16)$$

in terms of the known propagators for each of the two reactive sites— viz. the Green's functions $G_M(r, r'; t)$ and $G_P(r, r'; t)$. The auxiliary field $\psi(r)$ defines the "form" of the reaction zone and is a dimensionless function that is nonzero in a very narrow region near the reaction center. Its normalization,

$$v_c \equiv \int d^3r \, \psi(r),$$

is the "reaction volume."

For diffusional motion, the Green's functions obey the following equations:

$$\frac{\partial}{\partial t} G(r, r', t) = \hat{\mathcal{L}}(r) G,$$

with the initial condition

$$G(r, r', 0) = \delta(r - r')$$

and the reflecting boundary condition at contact. The operator $\hat{\mathcal{L}}(r)$ defines particle motion and in the case of free continuous diffusion takes the form

$$\hat{\mathcal{L}}(r) = D\nabla_r^2.$$

The solution of these Green's functions and the associated $p(t)$ (or $\tilde{p}(s)$) are known in a number of cases. For example, in the case of two diffusing noninteracting spherical particles [65],

$$\tilde{p}(s) = \frac{1}{4\pi D\sigma (1 + \sqrt{s\sigma^2/D})}. \quad (17)$$

The steady-state result, Eq. (14), can be obtained directly from Eq. (17) exploiting the formula $k_{d,r}(t \to \infty) = \lim_{s \to 0} s\tilde{k}_{d,r}(s)$.

In the present case of chain-growth polymerization, a Green's function, $G_P(r, r'; t)$, can be calculated for each addition of a monomer to the reactive group at the end of a polymer of degree n. Within the framework of the Rouse model, wherein a polymer chain is represented as a set of n beads connected by harmonic potentials, the propagator has the Gaussian form [10,66],

$$G_P(r, r' = 0; t) = (2\pi \phi_n(t))^{-3/2} \exp\left(-\frac{r^2}{2\phi_n(t)}\right), \quad (18)$$

where $\phi_n(t)$ is the mean-square displacement,

$$\phi_n(t) = 6D_n t + \frac{4nl^2}{\pi^2} \sum_{q=1}^{\infty} \frac{1 - \exp(-q^2 t/\tau_n)}{q^2}. \quad (19)$$

Here $\tau_n = n^2 L^2/(3\pi^2 D_1)$ are the Rouse relaxation times, D_1 is the diffusion coefficient of a monomer, and $D_n = D_1/n$ is that of the center of mass of a chain of degree n. Larger molecules will necessarily lead to longer relaxation times contributing to the time scales in the reaction rates. In another limiting case, when polymer chain configurations reach quasi-equilibrium sufficiently quickly (as compared with the long time scales of polymerization), the nonconstant behavior of the rate coefficients is caused by the changing density of solution. This alters the apparent diffusion of monomers, and reduces the rates (14) to:

$$k_{d,r}(t) = \frac{\kappa_{d,r}k_D(t)}{k_D(t) + \kappa_r}. \tag{20}$$

Even in dilute solutions, the chain relaxation effect, Eq. (19), included in Eq. (15), leads to time-dependent rates in Eq. (20) because it alters the diffusion coefficient of the polymer, D_P, in the course of polymerization, so that

$$D_P \equiv \lim_{t\to\infty} \phi_n(t)/(6t) = D_n.$$

In general, when both relaxational and environmental effects take place, they must be combined to construct the resulting equations for the time-dependent rates. Regardless of what is causing the time dependence in the rate coefficients, the CSE, Eq. (8), can be written in the form

$$\dot{n} = -h^2(t)\big(k_d(0) - k_r(0)[M]\big) + \sqrt{k_d(0) + k_r(0)[M]}h(t)\xi(t), \tag{21}$$

where $k_{d,r}(0)$ denote the initial values of the rate coefficients and the function $h(t) = \sqrt{k_{d(r)}(t)/k_{d(r)}(0)}$ defines their nonstationarity. If the rates are expressed by Eq. (20), then

$$h(t) = \sqrt{\frac{k_D(t)}{k_D(t) + \kappa_r}}.$$

The nonstationary function $h(t)$ is slowly varying to the extent that it effectively remains constant during any elementary reaction event.

3. POLYMERIZATION EVENTS MODELED BY LANGEVIN DYNAMICS

3.1 Nonstationary Langevin dynamics

Another promising way of extending the CSE to the nonequilibrium regime is to introduce a new effective coordinate, $R(t)$, which plays the role of the length of a polymer chain, so that

$$\langle R(t)\rangle = L\langle n(t)\rangle, \tag{22}$$

where L is the average monomer length. The "motion" along this variable is described by the generalized Langevin equation (GLE),

$$\ddot{R} = -\frac{\partial F(R)}{\partial R} - \int_0^t \gamma(t - t')\dot{R}\,dt' + \xi_{th}(t), \tag{23}$$

$$\langle \xi_{th}(t)\xi_{th}(t')\rangle = k_B T\gamma(t - t'),$$

where $\gamma(t - t')$ is the friction kernel, $F(R)$ represents the potential of mean force (PMF) and $\xi_{\text{th}}(t)$ is the stochastic Gaussian force. The parameters and functions in this equation are chosen in order to mimic the behavior of reaction processes that obey Eq. (22) on a long time scale. The qualitative properties and quantitative structure of the PMF is critically connected to this behavior, and they are discussed in detail in the next section.

Nonstationarity can be introduced into the GLE through an auxiliary function $g(t)$ so as to modulate the amplitude of the stochastic force [49,50]. The resulting equation of motion has been called the iGLE where the "i" refers to the irreversibly changing environment, and it takes the form,

$$\ddot{R} = -\frac{\partial F(R)}{\partial R} - \int_0^t g(t)\gamma(t - t')g(t')\dot{R}\,dt' + g(t)\xi_{\text{th}}(t). \tag{24}$$

The placement of $g(\cdot)$ on both sides of the friction kernel is necessary because the fluctuation-dissipation relation with modified forces,

$$\langle g(t)\xi_{\text{th}}(t)g(t')\xi_{\text{th}}(t')\rangle = k_{\text{B}}Tg(t)g(t')\gamma(t - t'),$$

must hold in the nonstationary environment [67,68]. Note that the nonstationary effects defined in Eq. (24) by the function $g(t)$ occur on a time scale that is usually much longer than the solvent relaxation time.

The growth of the effective polymers from an initial configuration of activated monomers is clearly visible in the time-dependent average of the polymers, $\langle R(t)\rangle$. The time-dependent distribution of polymer lengths can also be obtained from these calculations because the length of each of the polymers is known at a given time t. The iGLE model for polymerization has been seen to be capable of reproducing the rich structure of polymerization correctly in a couple examples of living polymerization as shown in Section 4.

3.2 The potential of mean force driving polymerizing events

Although the computational integration of the iGLE dynamics is compellingly simple, its use is predicated on the specification of the potential of mean force (PMF) and the nonstationary friction kernel. The original articulation—motivated heuristically using the method of steepest descents—of the PMF for polymer growth was specified as

$$e^{-\beta F(R)} \equiv Q^{-1} \sum_k \int_{\Omega_k} d\bar{\mathbf{r}} \exp(-\beta V(\bar{\mathbf{r}})) \delta\left\{ R - \sum_{i=1}^{k-1} |\mathbf{r}_{i+1} - \mathbf{r}_i| \right\}, \tag{25}$$

where $\beta = 1/k_{\text{B}}T$ is the inverse temperature; R is a coordinate corresponding not to the size (or end-to-end distance) of the polymer [69–71] but to its contour length [49]; R should be interpreted as the effective global reaction path coordinate for the chain polymerization. $V(\bar{\mathbf{r}})$ is the potential interaction between the n-mers represented by the $3k$-dimensional vector, $\bar{\mathbf{r}} \equiv (\mathbf{r}_1, \mathbf{r}_2, \ldots, \mathbf{r}_k)$, where $\mathbf{r}_i$ denotes the position of the ith monomer. Q is the partition function of the monomer. The choice of Q sets the zero of free energy to be at R near the average monomer length L. Notice that sum over the space Ω_k (which is the space of all phantom polymer chains with k monomer units) in addition to the δ-function constraint distinguishes this PMF from the usual polymer PMF [69–71] that characterizes the polymer size in a constant k ensemble.

The direct calculation of the PMF through Eq. (25) is extremely complicated even in the case of Gaussian chains and can be performed only numerically. However, the main features of this potential can be deduced from the common properties of the polymerization process. Each chain-growth reaction step which leads to an increase in the chain length, R, by L decreases the free energy by ΔG_L. This can be mimicked, minimally, by a simple biased potential,

$$F_b(R) = -f_b R, \tag{26}$$

where the slope $f_b = \Delta G_L / L$, complies with this condition. In order to further obtain a microscopic description of each association or dissociation reaction step, Eq. (26) can be modified by introducing wells at the typical lengths associated with each n-mer and barriers between adjoining n- and $(n + 1)$-polymers. Such a multiple well and barrier structure is known to lead to local activated rates between the wells from the theory of chemical reactions [72]. This phenomenological potential called the polymer growth potential (pgp) can be constructed based on a series of merged double well potentials with barrier heights and an external force f_b. Suppose the well frequency is ω_0 and the barrier frequency is ω_1, then the PMF can be written as:

$$F_{pgp}(R) \equiv \begin{cases} \frac{1}{2}\omega_0^2(R - kL)^2 - kf_b L \\ \quad \text{for } R < L + a_- \text{ with } k = 1 \quad \text{or} \quad kL - a_+ < R < kL + a_-, \\ -\frac{1}{2}\omega_1^2(R - kL - R_m)^2 - kf_b L + E_+^\dagger \\ \quad \text{if } kL + a_- < R < (k+1)L - a_+, \end{cases} \tag{27}$$

where R_m is the relative position of the potential maximum,

$$R_m = \frac{L}{2} - \frac{2f_b}{\omega_0^2} - \frac{2f_b}{\omega_1^2}, \tag{28}$$

$a_\pm$ define the sewing points of the piecewise continuous curve (27),

$$a_\pm = \frac{\omega_1^2 L/2}{\omega_0^2 + \omega_1^2} \pm \frac{2f_b}{\omega_0^2}, \tag{29}$$

and the barrier height in the forward reaction, $E_+^\dagger$, is expressed through the frequencies, ω_0 and ω_1, as

$$E_+^\dagger = \frac{(L - 4f_b(\omega_0^{-2} + \omega_1^{-2}))^2}{8(\omega_0^{-2} + \omega_1^{-2})} = \frac{R_m^2}{2(\omega_0^{-2} + \omega_1^{-2})}. \tag{30}$$

(The barrier height in the backward direction is $E_-^\dagger = E_+^\dagger + f_b L$.) The polymer growth PMF (pgp) is an extension of the double well model of Straub et al. [73].

iGLE model, the PMF is expressed by the barrier height which is the ratio of the activation energy over thermal energy. The activation energy for this polymerization process is about 20 kJ/mol and the deactivation energy is 29 kJ/mol [80]. The ratio of the barrier height over k_BT is about 9 and k_BT is in units of kJ/mol. The reference temperature, for which $k_BT_0 = 1$ kJ/mol, is 120 K.

With this assignment, numerical simulations of the iGLE can be compared to the behavior of corresponding physical systems with specified units. In the following text, we call the first set of data ($T_e = 267$ K) sample 1, and the second set of data ($T_e = 271$ K) sample 2 as shown in Table 10.1. The fraction of monomer remaining as a function of time for poly(α-methylstyrene) in THF initiated by sodium naphthalide is shown in Figure 10.2. The dots are the experimental data (sample 1) and the curve is the best fit of the theoretical model described in this work. The simulation results fit the experimental data very well with $\gamma_0 = 8$ and $\zeta = 0.95$. The equilibrium monomer concentration is reached in a relatively short amount of time and the extent of polymerization is 75%. For sample 2, we used the same γ_0 value because both of them operate in the same solvent-tetrahydrofuran. After a temperature quench from above the polymerization temperature to below the polymerization temperature, the polymerization process was initiated and the extent of polymerization is 62% at equilibrium at a temperature of 271 K. The high equilibrium temperature makes the free energy higher since entropy and enthalpy are all negative for this reaction. We adjust the value of ζ to account for the temperature difference. The result is shown in Figure 10.3. The temperature change

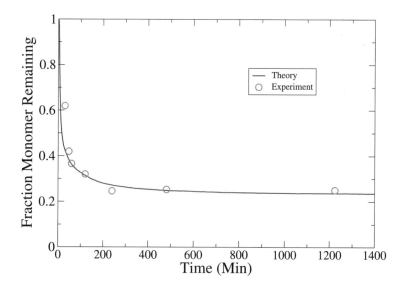

FIGURE 10.2   Experimental and theoretical fraction of monomers remaining as a function of time for poly(α-methylstyrene) in THF initiated by sodium naphthalide. The two sample batches are described in Table 10.1. The dots are the experimental data, and the solid curve is the theoretical model. The simulation parameters for batch sample 1 (dots) are: $N = 51,170$, $A = 128$, $T = 2.22T_0$, $E^\dagger = 19.98k_BT_0$, $f_b = 10$, $\zeta = 0.95$, $\gamma_0 = 8$.

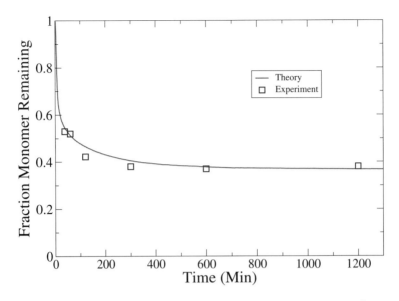

FIGURE 10.3   Experimental and theoretical fraction of monomers remaining as a function of time for poly(α-methylstyrene) in THF initiated by sodium naphthalide. The squares are experimental data and the solid curve is our theoretical model. The simulation parameters for Batch sample 2 (squares) are: $N = 51170$, $A = 123$, $T = 2.26T_0$, $E^\dagger = 19.98k_BT_0$, $f_b = 10$, $\zeta = 1.52$, $\gamma_0 = 8$.

also affects the solvent friction. If there is no reaction, viscosity decreases as temperature increases.

The results of the simulations shown above suggest that the parameter ζ should be related to the temperature. Greer has studied the extent of polymerization as a function of temperature for living poly(α-methylstyrene) in THF initiated by sodium naphthalide with mole fraction of monomers $x_m^0 = 0.15378$, and the mole ratio of initiators to monomers $r = 0.0044$. Based on their data, simulations were completed. The results are shown in Table 10.2. Thus we obtained the relation between ζ and temperature as shown in Figure 10.4. As temperature increases, the value of ζ increases monotonically and the extent of polymerization increases too.

4.3 Experimental system two: 4-vinylbenzocyclobutene

The polymerization of 4-vinylbenzocyclobutene in benzene using sec-butyllithium as the initiator at room temperature has recently been observed [82]. The number of average molecular weights at different times has been measured using size exclusion chromatography (SEC). These results are shown in the Table 10.3.

The various effective lengths, i_k, listed in Table 10.3 have been inferred from the experimental data using the following three different procedures:

(1) Suppose the measured average molecular weight at $t = 53$ min is accurate. For living polymerization, the number average molecular weight, M_n, is a linear function of conversion. (Column 3 of Table 10.3 also shows the recalcu-

Table 10.2   Batch samples of living poly(α-methylstyrene) in tetrahydrofuran, the table lists the data of temperature, barrier height, extent of conversion and the corresponding ζ value. The first and third columns are taken from [81]

Temperature (K)	Barrier height ($k_B T$)	Conversion	ζ
284.753	9.41	0.29 ± 0.02	4.8
282.350	9.487	0.41 ± 0.01	3.0
280.264	9.569	0.39 ± 0.01	3.2
278.317	9.623	0.54 ± 0.01	1.9
275.707	9.715	0.55 ± 0.01	1.85
273.095	9.808	0.668 ± 0.006	1.22
267.789	10.00	0.749 ± 0.004	0.92
263.480	10.165	0.800 ± 0.003	0.75
258.792	10.350	0.858 ± 0.003	0.55
254.677	10.520	0.894 ± 0.003	0.45

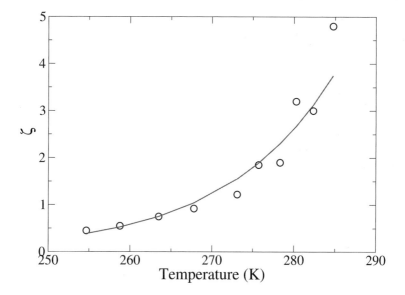

FIGURE 10.4   Extent of polymerization as a function of temperature for living poly(α-methylstyrene) in THF initiated by sodium naphthalide with $x_m^0 = 0.15378, r = 0.0044$. The solid curve is the exponential fit. The corresponding simulation parameters are: $A = 204$, $N = 46{,}288, f_b = 10, E^\dagger = 22.2 k_B T_0$.

lated/corrected M_n in parentheses.) Since the molecular weight of a monomer is 130 g/mol, the polymer length i_1 can be calculated by using $i_1 = M_n/130$.

Table 10.3 Polymerization of 4-vinylbenzocyclobutene in benzene using sec-butyllithium as the initiator at room temperature. ϕ is the extent of polymerization, M_n is the number average molecular weight, i_1, i_2 and i_3 are the polymer lengths obtained using different experimental data. The first three columns are taken from [82]

Time (min)	ϕ	M_n (corrected)	i_1	i_2	i_3
53	10%	4100	31.5	31.2	40.02
90	18%	6200 (7380)	56.77	56.16	64.87
182	33%	12,100 (13,530)	104.08	102.96	117.27
470	71%	24,300 (29,100)	223.92	221.52	219.64

(2) Based on the concentration of initiators and monomers, the ratio of monomers and initiators can be found:

$$[M_0]/[I_0] = 311.68. \tag{34}$$

This value corresponds to the polymer length when conversion is 100%. The polymer length i_2 is calculated using the conversion factor multiplied by this maximum length.

(3) According to Figure 3 in [82], the apparent rate constant is $k_{app} = 0.00259$ min^{-1}. Assuming that the reaction follows first-order kinetics, the free monomer concentration $[M_t]$ can be approximately found at different times by use of the formula

$$\ln([M_0]/[M_t]) = k_{app}t.$$

Calculating the conversion factor, $1 - [M_t]/[M_0]$, and multiplying it by the maximum length, 312, one then obtains the polymer length i_3. For the time points taken from Table 10.3, the results of these calculations are shown in Table 10.4 The length obtained is larger than that using the other two methods shown in Table 10.3.

In the following simulations, the initial concentrations of monomers and initiators were taken equal to $[M]_0 = 0.24$ mol/l and $[I]_0 = 0.00077$ mol/l, correspondingly, in accordance with relation (34). For the system volume equated to $3.0 \cdot 10^8$ Å^3, the number of activated monomers is $A = 139$, and the total number of monomers is $N = 43,344$.

Compared to poly(α-methylstyrene), this polymerization reaction is irreversible and there are few monomers left in solution at the end. To test our model, we use the same parameter values and turn off the back reaction to simulate the irreversible reaction. As shown in Figure 10.5, the polymer grows faster for an irreversible reaction as expected.

In order to completely specify the model, there is one remaining unknown parameter for this system. In particular, the activation energy for the anionic polymerization of 4-vinylbenzocyclobutene is not available. Here we use two different values for the activation energy (20 kJ/mol and 63 kJ/mol) to model the properties of this living polymerization reaction. These are the activation energies for

Table 10.4

Time (min)	$[M_t]$ (mol/l)	$1 - [M_t]/[M_0]$	i_3
53	0.21	12.83%	40.02
90	0.19	20.80%	64.87
182	0.15	37.59%	117.27
470	0.07	70.40%	219.64

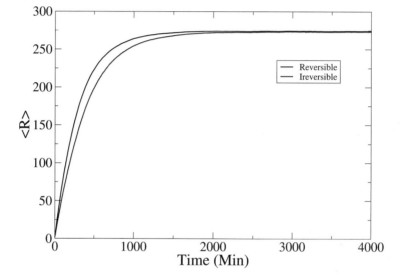

FIGURE 10.5   The average polymer length is displayed as a function of time for reversible (black) and irreversible (dark gray in print and blue on-line) polymerization. The simulation parameters are: $A = 139, N = 43{,}344, \zeta = 0.5, \gamma_0 = 180$. The irreversible case rises more slowly and is obtained by turning off the back reaction.

α-methylstyrene [81] and styrene [83], respectively. The use of these values is motivated by the fact that 4-vinylbenzocyclobutene and α-methylstyrene can be considered as derivatives of styrene. The polymer length is calculated as a function of time at two different barrier heights, as shown in Figure 10.6. When using a low barrier height, a linear first-order time-conversion kinetics is seen. When the barrier height is as high as $24.3k_BT$, the polymerization rate is slow at the beginning and then speeds up, showing an "S" shape behavior. This can be seen more clearly in Figure 10.7.

The SEC results suggest that the number average molecular weight increases with the polymerization time. The distribution becomes narrower and narrower, which implies that the polymer is approaching a uniform length. Our length distribution (Figure 10.8) indicates this as well. It is difficult to obtain the same SEC curve because we do not know the calibration curve of the polystyrene standard

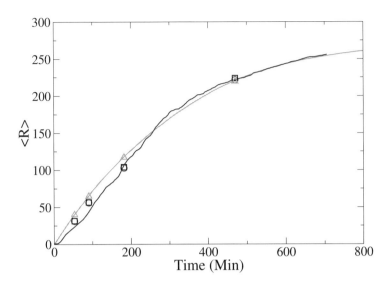

FIGURE 10.6 Theoretical and experimental polymer length as a function of time for the irreversible polymerization with different barrier height. The blue—or black in grayscale—curve is the simulation result using $A = 139$, $N = 43{,}344$, $f_B = 30$, $\gamma_0 = 56$, $E^{\dagger} = 24.3 k_B T$, $\zeta = 0.16$; the green—or light gray in grayscale—curve is the simulation result using $A = 139$, $N = 43{,}344$, $f_B = 10$, $\gamma_0 = 75$, $E^{\dagger} = 8.0 k_B T$, $\zeta = 0.5$. The dots, squares and triangles correspond to the polymer length calculated using the concentration, M_n and rate constant.

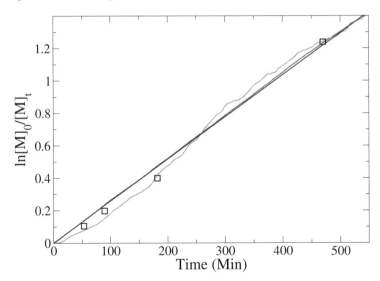

FIGURE 10.7 The kinetics of the anionic polymerization of 4-vinylbenzocyclobutene using s-BuLi as initiator in benzene at 25 degrees. The squares represent the experimental SEC results, the green—or gray—curve is the simulation result using $f_B = 30$, $\gamma_0 = 56$, $E^{\dagger} = 24.3 k_B T$, $\zeta = 0.16$, the red—or dark gray in grayscale—curve is the simulation result using $f_B = 10$, $\gamma_0 = 75$, $E^{\dagger} = 8.0 k_B T$, $\zeta = 0.5$, and the solid blue—or black in grayscale—curve is first-order kinetics.

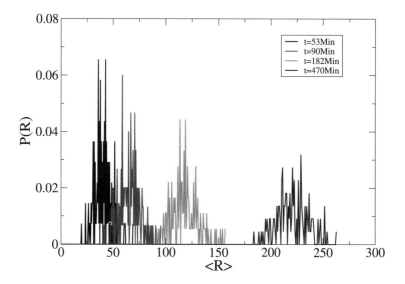

FIGURE 10.8 Polymer length distribution of poly(4-vinylbenzocyclobutene) at different time with $A = 139, N = 43,344, \zeta = 0.5, \gamma_0 = 75, f_b = 10, E^\dagger = 8k_B T$.

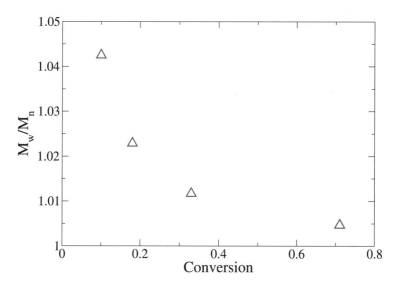

FIGURE 10.9 Polydispersity index (PDI) of poly(4-vinylbenzocyclobutene)as a function of conversion with $E^\dagger = 8k_B T$. The PDI value decreases with conversion and is close to 1.

under the current experimental conditions. A good way to link the experimental distribution and theory is to calculate the polydispersity index (PDI). As shown in Figures 10.9 and 10.10, we obtain different PDI values with different barrier heights. By using a low barrier height, the PDI value is close to 1. This is consistent with the experimental data.

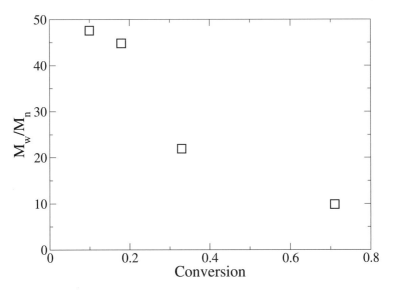

FIGURE 10.10   Polydispersity index (PDI) of poly(4-vinylbenzocyclobutene) as a function of conversion with $E^{\dagger} = 24.3k_{\mathrm{B}}T$. The PDI value decrease with conversion and it is between 10–50.

5. DISCUSSION

The modified iGLE model has been applied to two experimental systems: α-methylstyrene and 4-vinylbenzocyclobutene. Although the polymerization dynamics of these systems vary because of different reagent and operational conditions, the qualitative features of the dynamics between both the experimental observations and the simulations appear to be in good agreement. The theoretical model also appears to give rise to qualitative empirical behaviors known to occur in other polymerizing systems. For example, as shown in Figures 10.6 and 10.7, the simulations of the theoretical model at high enough barriers even exhibits the Trommsdorff effect [84] in which there exists an auto-acceleration at the onset of polymerization.

To test the generality of the iGLE approach, an irreversibly polymerizing system, 4-vinylbenzocyclobutene, was chosen. Since there is no data available for the activation energy of the polymerizing chain-growth steps, two different possible limits for the activation energy were employed. In either case, the remaining parameters of the iGLE could be adjusted to fit to the experimental data that is available. While it is reassuring that the experiments can be described within this formalism, it does point to the need for more detailed time-dependent data for these processes in order to be able to specify the model more precisely.

By studying the temperature effect on the polymerization of α-methylstyrene, the relation between the model parameter ζ and temperature has been obtained. Knowledge of this parameter allows for further calculations of the polymerization dynamics at different temperatures. Moreover it suggests an interrelation between the average size of the polymer distribution and the reaction coordinate between the activated ends and a given monomer. The former defines the density of polymer solvent which in turn affects the accessibility of monomers to the reaction sites. A challenge to future work is the exact determination of this relation either from specific microscopic interactions, or perhaps even from some universal properties.

The polymer-length distributions and PDI values have also been obtained. Under the quasi-equilibrium conditions arising from the separation of time scales described above, these distributions have the structure of a Flory distribution [1]. Through fits of the iGLE distribution with the Flory distribution, the time-dependent extent of conversion may thus be obtained [49]. If the activation energy is high, it takes particles more time to cross the self-similar barriers with an average time that is near the same mean first passage time. Thus it is possible to get a very narrow distribution if the simulation is long enough to exhaust nearly all monomers, and the polymer formation is unaffected by termination and transfer reactions. Thus, just as in living polymerization, the iGLE theory leads to polymer-growth computational phenomena that initially exhibits narrow polymer-length distributions.

6. CONCLUSION

In this review, we have emphasized two stochastic methods suitable for simulating polymerizing systems with specific reference to those of involving living and thermosetting polymerization. In both of these systems the environment is changing dramatically with the progress of the underlying microscopic polymerizing reactions, and hence any theory for these systems must necessarily include a *nonstationary* description of the solvation of the effective reaction coordinate. Although the use of this theory has been formally described earlier for the living and thermosetting systems [49,50], the present work provides explicit applications to two distinct living polymerizing systems only.

In our first approach, we have modified the CSE method so as to include nonstationary in the dissipative terms. Although this is a fruitful approach to account for many different types of nonstationarities, it suffers from the fact that said accounting is not specific in a single unified way. Moreover, the theory requires one to calculate or compute direct and inverse Laplace transforms of initially unknown terms so as to include some of the effects correctly. Though a self-consistent iterative procedure may ultimately lead to success, at present, this approach appears not to be the right choice. Nevertheless, the formalism has been of use in providing a better understanding of the iGLE approach.

In our second approach, we use a nonstationary version of the GLE—the so-called iGLE [67]—to propagate an effective polymer length that plays the role of the reaction coordinate. The forces in this dissipative equation-of-motion are surmised by a PMF which combines the properties of the double-well potential for forward and backward reactions within one biased periodic potential. The amplitudes of the friction and stochastic forces account for the changing environment and define the "diffusion" properties along the reaction path. Using the iGLE formalism to model polymerization provides a bridge between simple kinetics and full dynamics. The new theory reduces to the nonstationary CSE method under steady-state conditions, and the CSE in its turn produces the kinetic equations at homogeneous stationary conditions after the averaging over fluctuations.

ACKNOWLEDGMENTS

This work was partly supported through unrestricted grants by the US National Science Foundation, the Alexander von Humboldt-Foundation, and the Rohm & Haas Company. The early stages of this work were supported by NSF Grant CHE 02-123320, and the computational facilities at the CCMST have been supported under NSF Grant CHE 04-43564.

REFERENCES

1. Flory, P.J. Molecular size distribution in three dimensional polymers. I. Gelation. J. Am. Chem. Soc. 1941, 63, 3083.
2. Stockmayer, W.H. Theory of molecular size distribution and gel formation in branched-chain polymers. J. Chem. Phys. 1943, 11, 45.
3. Flory, P.J. Principles of Polymer Chemistry. Ithaca, NY: Cornell Univ. Press; 1953.
4. Tobolsky, A.V., Eisenberg, A. Transition phenomena in equilibrium polymerization. J. Colloid Sci. 1962, 17, 49.
5. Scott, R.L. Phase equilibria in solutions of liquid sulfur. I. Theory. J. Phys. Chem. 1965, 69, 261.
6. Flory, P.J. Statistical Mechanics of Chain Molecules. New York: Wiley-Interscience; 1969.
7. de Gennes, P.G. Exponents for the excluded volume problem as derived by the Wilson method. Phys. Lett. A 1972, 38, 339.
8. des Cloiseaux, J. The Lagrangian theory of polymer solutions at intermediate concentrations. J. Phys. (Paris) 1975, 36, 281.
9. de Gennes, P.G. Scaling Concepts in Polymer Physics. Ithaca, NY: Cornell Univ. Press; 1979.
10. Doi, M., Edwards, S.F. The Theory of Polymer Dynamics. Oxford: Clarendon; 1986.
11. Freed, K.F. Renormalization Group Theory of Macromolecules. New York: Wiley-Interscience; 1987.
12. Wheeler, J.C., Pfeuty, P. The $n \to 0$ vector model and equilibrium polymerization. Phys. Rev. A 1981, 24, 1050.
13. Schaefer, L. Chain-length distribution in a model of equilibrium polymerization. Phys. Rev. B 1992, 46, 6061.
14. O'Shaughnessy, B. From mean field to diffusion-controlled kinetics: Concentration-induced transition in reacting polymer solutions. Phys. Rev. Lett. 1993, 71, 3331.
15. O'Shaughnessy, B., Yu, J. Autoacceleration in free radical polymerization. Phys. Rev. Lett. 1994, 73, 1723.
16. Tobita, H. General matrix formula for the weight-average molecular weights of cross-linked polymer systems. J. Polym. Sci. B Polym. Phys. 1998, 36, 2423.
17. Kennedy, S.J., Wheeler, J.C. Critical and tricritical phenomena in "living" polymers. J. Chem. Phys. 1983, 78, 953.
18. Corrales, L.R., Wheeler, J.C. Tetracritical and novel tricritical points in sulfur solutions: A Flory model for polymerization of rings and chains in solvent. J. Chem. Phys. 1989, 90, 5030.
19. Zheng, K.M., Greer, S.C., Corrales, L.R., Ruiz-Garcia, J. Living poly(α-methylstyrene) near the polymerization line: II. Phase diagram in methylcyclohexane. J. Chem. Phys. 1993, 98, 9873.
20. Milchev, A. Phase transitions in polydisperse polymer melts. Polymer 1993, 34, 362.
21. Milchev, A., Landau, D.P. Monte Carlo study of semiflexible living polymers. Phys. Rev. E 1995, 52, 6431.

22. Szwarc, M. Carbanions, Living Polymers and Electron Transfer Processes. New York: Wiley-Interscience; 1968.
23. Szwarc, M., Van Beylen, M. Ionic Polymerization and Living Polymers. New York: Chapman & Hall; 1993.
24. Greer, S.C. Living polymers. Adv. Chem. Phys. 1996, 94, 261.
25. Greer, S. Living polymers. Compl. Matls. Sci. 1995, 4, 334.
26. Milchev, A., Rouault, Y., Landau, D. Monomer-mediated relaxation in living polymers. Phys. Rev. E 1997, 56, 1946.
27. van der Schoot, P. Growth of living polymers in a good solvent. Europhys. Lett. 1997, 39, 25.
28. Rouault, Y. Living polymers in random media: A 2d Monte Carlo investigation on a square lattice. Eur. Phys. J. B 1998, 2, 483.
29. Wittmer, J., Milchev, A., Cates, M. Dynamical Monte Carlo study of equilibrium polymers: Static properties. J. Chem. Phys. 1998, 109, 834.
30. Wittmer, J.P., Milchev, A., Cates, M.E. Computational confirmation of scaling predictions for equilibrium polymers. Europhys. Lett. 1998, 41, 291.
31. Menon, G., Pandit, R. Crystallization and vitrification of semiflexible living polymers: A lattice model. Phys. Rev. E 1999, 59, 787.
32. Sarkar Das, S., Zhuang, J., Ploplis Andrews, A., Greer, S.C., Guttman, C., Blair, W. Living poly(α-methylstyrene) near the polymerization line. VII. Molecular weight distribution in a good solvent. J. Chem. Phys. 1999, 111, 9406.
33. Dudowicz, J., Freed, K.F., Douglas, J. Lattice model of living polymerization. I. Basic thermodynamic properties. J. Chem. Phys. 1999, 111, 7116.
34. Dudowicz, J., Freed, K.F., Douglas, J. Lattice model of living polymerization. II. Interplay between polymerization and phase stability. J. Chem. Phys. 2000, 112, 1002.
35. Dudowicz, J., Freed, K.F., Douglas, J. Lattice model of living polymerization. III. Evidence for particle clustering from phase separation properties and "rounding" of the dynamical clustering transition. J. Chem. Phys. 2000, 113, 434.
36. Pendyala, K., Gu, X., Andrews, K., Gruner, K., Jacobs, D.T., Greer, S.C. Living poly(α-methylstyrene) near the polymerization line. VIII. Mass density, viscosity, and surface tension in tetrahydrofuran. J. Chem. Phys. 2001, 114, 4312.
37. Niranjan, P.S., Forbes, J.G., Greer, S.C., Dudowicz, J., Freed, K., Douglas, J. Thermodynamic regulation of actin polymerization. J. Chem. Phys. 2001, 114, 10573.
38. Dudowicz, J., Freed, K.F., Douglas, J. Lattice model of living polymerization. IV. Influence of activation, chemical initiation, chain scission and fusion, and chain stiffness on polymerization and phase separation. J. Chem. Phys. 2003, 119, 12645.
39. Niranjan, P.S., Yim, P.B., Forbes, J.G., Greer, S.C., Dudowicz, J., Freed, K., Douglas, J. The polymerization of actin: Thermodynamics near the polymerization line. J. Chem. Phys. 2003, 119, 4070.
40. Bawn, C.E.H. The Chemistry of High Polymers. New York: Wiley-Interscience; 1948.
41. Koenig, J.L. Chemical Microstructure of Polymer Chains. New York: Wiley-Interscience; 1980.
42. Dušek, K. Cross-linking of epoxy resins. In: Riew, C.K., Gillham, J.L., editors. Rubber-Modified Thermoset Resins. Adv. Chemistry, vol. 208. Washington, DC: Am. Chem. Soc.; 1984.
43. Williams, R.J.J., Borrajo, J., Adabbo, H.E., Rojas, A.J. A model for phase separation during a thermoset polymerization. In: Riew, C.K., Gillham, J.L., editors. Rubber-Modified Thermoset Resins. Adv. Chemistry, vol. 208. Washington, DC: Am. Chem. Soc.; 1984, p. 195.
44. Dušek, K., Matějka, L. Rubber–toughened plastics. In: Riew, C.K., editor. Rubber–toughened plastics. Adv. Chemistry, vol. 222. Washington, DC: Am. Chem. Soc.; 1989, p. 303.
45. Gum, W.F., Riese, W., Ulrich, H., editors. Reaction Polymers. New York: Oxford Univ. Press; 1992.
46. Prepelka, P.J., Wharton, J.L. Reaction injection molding in the automotive industry. J. Cell. Plast. 1975, 11, 87.
47. Lee, L.J. Polyurethane reaction injection molding: Process, materials, and properties. Rubber Chem. Technol. 1980, 53, 542.
48. Castro, J.M., Lipshitz, S.D., Macosko, C.W. Laminar tube flow with a Thermosetting polymerization. AIChE J. 1982, 26, 973.
49. Hernandez, R., Somer, F.L. Stochastic dynamics in irreversible nonequilibrium environments. 2. A model for thermosetting polymerization. J. Phys. Chem. B 1999, 103, 1070.

50. Somer, F.L., Hernandez, R. Stochastic dynamics in irreversible nonequilibrium environments. 4. Self-consistent coupling in heterogeneous environments. J. Phys. Chem. B 2000, 104, 3456.
51. Hernandez, R., Somer, F.L. Nonstationary stochastic dynamics and applications to chemical physics. In: Schwartz, S.D., editor. Theoretical Methods in Condensed Phase Chemistry. The Netherlands: Kluwer Academic; 2000, pp. 91–116.
52. Vogt, M., Hernandez, R. A two-dimensional polymer growth model. J. Chem. Phys. 2001, 115, 1575.
53. Vogt, M., Hernandez, R. A three-dimensional polymer growth model. J. Chem. Phys. 2002, 116, 10485.
54. Gillespie, D.T. The chemical Langevin equation. J. Chem. Phys. 2000, 113, 297.
55. Gillespie, D.T. Stochastic simulation of chemical kinetics. Annu. Rev. Phys. Chem. 2007, 58, 35.
56. Risken, H. The Fokker–Planck Equation. New York: Springer-Verlag; 1989.
57. Agmon, N., Szabo, A. Theory of reversible diffusion-influenced reactions. J. Chem. Phys. 1990, 92, 5270.
58. Szabo, A. Theoretical approaches to reversible diffusion-influenced reactions—Monomer excimer kinetics. J. Chem. Phys. 1991, 95, 2481.
59. Sung, J., Lee, S. Relations among the modern theories of diffusion-influenced reactions. I. Reduced distribution function theory versus memory function theory of Yang, Lee, and Shin. J. Chem. Phys. 1999, 111, 10159.
60. Gopich, I.V., Szabo, A. Kinetics of reversible diffusion influenced reactions: The self-consistent relaxation time approximation. J. Chem. Phys. 2002, 117, 507.
61. Doktorov, A.B., Kipriyanov, A.A. A many-particle derivation of the Integral Encounter Theory non-Markovian kinetic equations of the reversible reaction A + B reversible arrow C in solutions. Physica A 2003, 319, 253.
62. Agmon, N., Popov, A.V. Unified theory of reversible target reactions. J. Chem. Phys. 2003, 119, 6680.
63. Agmon, N., Popov, A.V. Accurate solution for the many-body ABCD problem. Physica A 2003, 330, 150.
64. Popov, A.V., Agmon, N., Gopich, I.V., Szabo, A. Influence of diffusion on the kinetics of excited-state association-dissociation reactions: Comparison of theory and simulation. J. Chem. Phys. 2004, 120, 6111.
65. Rice, S.A. Diffusion-Limited Reactions, Comprehensive Chemical Kinetics, vol. 25. Amsterdam: Elsevier; 1985.
66. Park, P.J., Lee, S. Diffusion-influenced reversible energy transfer reactions between polymers. J. Chem. Phys. 2001, 115, 9594.
67. Hernandez, R., Somer, F.L. Stochastic dynamics in irreversible nonequilibrium environments. 1. The fluctuation-dissipation relation. J. Phys. Chem. B 1999, 103, 1064.
68. Hernandez, R. The projection of a mechanical system onto the irreversible Generalized Langevin Equation (iGLE). J. Chem. Phys. 1999, 111, 7701.
69. Edwards, S.F. The statistical mechanics of polymers with excluded volume. Proc. Phys. Soc. London 1965, 85, 613.
70. Edwards, S.F. The statistical mechanics of a single polymer chain. In: Proc. Nat. Bur. Stand. Conf. on Critical Points, vol. 273. Washington: National Bureau of Standards; 1965, p. 225.
71. Freed, K.F. Functional integrals and polymer statistics. Adv. Chem. Phys. 1972, 22, 1.
72. Steinfeld, J.I., Francisco, J.S., Hase, W.L. Chemical Kinetics and Dynamics. second ed. Upper Saddle River, NJ: Prentice Hall; 1999.
73. Straub, J.E., Borkovec, M., Berne, B.J. Non-Markovian activated rate processes: Comparison of current theories with numerical simulation data. J. Chem. Phys. 1986, 84, 1788.
74. Reimann, P., den Broeck, C.V., Linke, H., Hnggi, P., Rubi, J.M., Prez-Madrid, A. Giant acceleration of free diffusion by use of tilted periodic potentials. Phys. Rev. Lett. 2001, 87, 010602.
75. Hayashi, K., Sasa, S.I. Effective temperature in nonequilibrium steady states of Langevin systems with a tilted periodic potential. Phys. Rev. E 2004, 69, 066119.
76. Speck, T., Seifert, U. Restoring a fluctuation-dissipation theorem in a nonequilibrium steady state. Europhys. Lett. 2006, 74, 391.
77. Sakaguchi, H. Generalized Einstein relation for Brownian motion in tilted periodic potential. J. Phys. Soc. Jap. 2006, 75, 124006.
78. Sakaguchi, H. Efficiency and fluctuation in tight-coupling model of molecular motor. J. Phys. Soc. Jap. 2006, 75, 063001.

79. Zhuang, J., Sarkar Das, S., Nowakowski, M.D., Greer, S.C. Living poly(α-methylstyrene) near the polymerization Line. 6. Chemical kinetics. Physica A 1997, 244, 522.
80. Kalnin'sh, K.K. Polymerization/Depolymerization mechanism for "living" sodium poly-α-Methylstyryl and sodium polystyryl. Russian J. Appl. Chem. 2001, 74, 1913.
81. Sarkar Das, S., Ploplis Andrews, A., Greer, S.C. Living poly(α-methylstyrene) near the polymerization line: IV. Extent of polymerization as a function of temperature. J. Chem. Phys. 1995, 102, 2951.
82. Sakellariou, G., Baskaran, D., Hadjichristidis, N., Mays, J.W. Well-defined poly(4-vinylbenzocyclobutene): Synthesis by living anionic polymerization and characterization. Macromolecules 2006, 39, 3525.
83. Guerrero-Sanchez, C., Abeln, C., Schubert, U.S. Automated parallel anionic polymerizations: Enhancing the possibilities of a widely used technique in polymer synthesis. J. Polym. Sci. A Polym. Chem. 2005, 43, 4151.
84. Trommsdorff, E., Köhle, H., Lagally, P. Zur Polymerisation des Methacrylsäure-methylesters. Makromol. Chem. 1948, 1, 169.

Section 5
Emerging Technologies

Section Editor: Wendy Cornell

Merck Research Laboratories
P.O. Box 2000
Rahway, NJ 07065
USA

CHAPTER 11

SAR Knowledge Bases in Drug Discovery

Stefan Senger[*,1] and **Andrew R. Leach**[*]

Contents

1. INTRODUCTION

In recent years we have witnessed an explosion in the volume of data describing the interactions between small, drug-like molecules and biological targets. The introduction of automated methods for the synthesis and testing of compounds is one reason for this growth; a typical compound collection at a large pharmaceutical company now numbers at least one million samples and these can be tested using current high-throughput screening techniques on timescales measured in days. A second reason can be attributed to the growth in the number of organisations with the ability to generate such data; this now includes not only the commercial pharmaceutical sector but also groups funded through the public

[*] Medicines Research Centre, GlaxoSmithKline Research & Development, Gunnels Wood Road, Stevenage, Hertfordshire SG1 2NY, UK
E-mail: Stefan.x.senger@gsk.com (S. Senger)
[1] Corresponding author.

Annual Reports in Computational Chemistry, Vol. 4
ISSN 1574-1400, DOI: 10.1016/S1574-1400(08)00011-X

purse or by charities. There has also been a revolution in the mechanisms available for the publication of results and data, particularly via the Internet, compounded by a steady growth in the number of journals devoted to drug discovery.

With this data explosion comes the need for electronic mechanisms to store, search and retrieve relevant information for subsequent analysis—and hopefully beneficial insights. The relatively new development that is the primary subject of this review is the availability of databases (which we term "SAR knowledge bases") that contain large amounts of quantitative data characterising the activity of compounds in biological assays. Some of these databases are freely available over the web; others operate on a commercial basis. Some aim to be "comprehensive"; others limit themselves to a particular type of data, molecule, or biological target. Our main focus will be on curated databases that contain biological activity data linked to chemical structures, taken from published sources (including electronic publications) though we will also indicate some of the other sources that may be of interest. By no means is ours the first summary of the field; the interested reader is directed towards other reviews that provide a complementary perspective [1–3]. Moreover, it is important to recognise that the number of databases containing chemical and biological information is growing extremely rapidly and so even with our restricted focus it is inevitable that we can only provide a partial picture, both in terms of the databases themselves and their content. Our review is structured as follows. First, we summarise the key features and capabilities of the main databases currently available. We then indicate some of the ways in which the information in such databases can be used in drug discovery. Finally, we consider possible future directions.

2. OVERVIEW OF SAR KNOWLEDGE BASES

2.1 Databases containing "raw" screening data

As indicated above, the focus of this review is on curated databases containing chemical structure and biological activity taken from literature sources. We nevertheless recognise the existence of a growing number of databases that also contain such structure and activity data but which do not fall within our primary scope. Of particular relevance here are databases that contain "raw" screening data, often generated by publicly-funded research initiatives. Two prominent examples are Pubchem [4] and Chembank [5].

Pubchem (http://pubchem.ncbi.nlm.nih.gov) is one of the many databases hosted by the National Centre for Biotechnology Information (NCBI) at the US National Institutes of Health (NIH) and is a key component of the NIH Roadmap Initiative on molecular libraries [6] with its focus "on the chemical, structural and biological properties of small molecules, particularly their application as diagnostic and therapeutic agents". It contains more than 19 million unique structures and more than 40 million substances or samples (a given chemical structure may occur in more than one substance). The biological data in PubChem is obtained from a variety of sources, prominent among which are the ten centres that comprise the

NIH Molecular Libraries Screening Center Network (MLSCN) [7]; currently in excess of 750,000 substances have biological data from at least one of the more than 1000 assays listed. The primary aim of the MLSCN is not to develop drugs but to identify small-molecule chemical probes for basic research; as a consequence the MLSCN screening libraries may contain some molecules that would be considered "unacceptable" (i.e. not "drug-like" or "lead-like") by pharmaceutical companies. The data are openly accessible in PubChem immediately available after deposition. In addition to primary screening results, PubChem acts as a repository for results generated as part of any efforts to follow up hits of interest (e.g. secondary assay data, results from follow-up compound libraries, and synthetic protocols). Pubchem links via the NCBI's Entrez system to other databases such as the PubMed (which contains citations, many with abstracts), PubMed Central (which contains full-text articles) and various information sources about the biological targets.

The origins of ChemBank (http://chembank.broad.harvard.edu) lie in a collaboration between the National Cancer Institute (its Initiative for Chemical Genetics) and the Harvard Institute of Chemistry and Cell Biology; it is now hosted at the Chemical Biology Program and Platform at the Broad Institute of Harvard and MIT. Uniquely among the publicly-funded efforts it contains raw screening data primarily from that one institute and has a tightly defined structure with regard to the way in which the information is organised. As such it is perhaps more akin to the screening databases contained within the commercial pharmaceutical sector. PubChem by contrast compiles data from a number of sources and relies upon the information and interpretation provided by the submitting organisation. Another difference is ChemBank's embargo policy; newly generated data are only available to those scientists who have deposited compounds or performed screening experiments at the two institutions for a one year period. ChemBank contains more than 1.2 million chemical structures with data from more than 2500 high-throughput assays and approximately 90 small-molecule microarray assays. Detailed information is provided, including plate locations; ChemBank also provides a number of tools for the visualisation and analysis of this data.

Both Pubchem and ChemBank are publicly available over the Internet; as such they can be considered successors to the databases hosted at the National Cancer Institute (NCI). The NCI data set contains more than 250,000 molecules together with biological activity obtained from their anti-cancer and HIV screens. For many years this data set was the only one of any significant size available publicly and as such it has been widely used to test new data-mining approaches [8].

2.2 Databases of drugs and drug candidates

Information on drug candidates and marketed drugs has been available for many years on a commercial basis via data sources such as Pharmaprojects (http://www.pharmaprojects.com), the World Drug Index from Thompson Scientific (http://scientific.thompson.com) and the MDL Drug Data Report (MDDR) (http://www.mdl.com). These databases contain a significant amount of information but the activity data is reported as activity classes (e.g. "Angiotensin II Antagonist";

"Thrombin inhibitor"; "Antidiabetic") rather than quantitative activity against a defined biological target. Nevertheless, they have been very widely used for the development and assessment of chemoinformatic data-mining techniques. The focus of these databases means that much of their information comes from sources other than the primary scientific literature such as patents and conference proceedings. Other databases that contain information useful in drug discovery include the Comprehensive Medicinal Chemistry database (CMC, also from MDL) and the MedChem database (http://www.daylight.com), both of which have an emphasis on the physicochemical properties of molecules.

More recently, efforts have been initiated to provide in a publicly-accessible manner chemical and biological information on drug molecules and their targets, leading to a number of databases such as ChEBI [9] (http://www.ebi.ac.uk/chebi) and DrugBank [10,11] (http://www.drugbank.ca). The latter in particular is a widely-used resource that contains detailed information on approximately 4900 drug molecules including all drugs approved in North America, Europe and Asia. For each entry DrugBank may contain up to 100 different pieces of data related to its chemical structure, properties, biological activity, pharmacology, mechanism of action and clinical information including toxicity and pharmacokinetics. This data is drawn from many sources including textbooks, the scientific literature and on-line resources; key data is manually inspected and checked by relevant experts. This attention to detail and its attempt to achieve both "breadth" and "depth" of coverage has led to its use in a variety of areas [11] such as in silico drug discovery, metabolism prediction, target prediction as well as more general education.

2.3 Databases for structure-based design

Advances in structural biology have resulted in an exponential increase in the number of proteins and protein–ligand complexes for which detailed 3-dimensional information is available, typically from X-ray crystallography or NMR. This has been accompanied by a continual desire to use structural information in drug design. Three general applications can be identified [12]:

(1) predicting the binding mode of a known active ligand;
(2) identifying new ligands using virtual screening;
(3) predicting the binding affinities of related compounds from a known active series (commonly referred to as "scoring").

As the Protein Databank (PDB) has grown so it has been recognised that there is a need for tools to analyse and interrogate this data. Some of these tools are provided by the three international organisations that host the PDB (the Research Collaboratory for Structural Bioinformatics in the US, http://www.rcsb.org, the European Bioinformatics Institute in Europe, http://www.ebi.ac.uk, and the Protein Data Bank Japan, http://www.pdbj.org). Other tools have been developed in academic laboratories and software companies. Of particular relevance to this review are the various efforts to compile protein–ligand activity data for those systems where structural information is available. The resulting databases represent a useful resource, particularly for those research groups attempting to develop new scoring algorithms. As a result of these compilation efforts a number of such databases are now available over the Internet, including PDB-Bind [13–15], BindingMOAD [16,17], BindingDB [18–21], the Protein Ligand Database [22] (http://lpdb.chem.lsa.umich.edu), the Ligand–Protein DataBase [23] (http://www-mitchell.ch.cam.ac.uk/pld) and AffinDB [24]. Attention here will be focussed on the first three of these as they contain the largest volumes of data.

PDBBind (http://www.pdbbind.org) and BindingMOAD (MOAD standing for Mother Of All Databases; http://www.bindingmoad.org) have a very similar aim, which is to annotate protein–ligand complexes from the PDB with binding or activity data (indeed, these databases originate from the same institution). Both groups adopt a similar approach to processing the data, whereby the initial set of complexes from the PDB is subjected to a series of filters that aim to identify those complexes of specific interest (typically those containing some form of drug-like molecule). The literature was then examined to extract the corresponding binding data. Due to variations in the criteria that the two groups use the number of entries is somewhat different; PDBBind currently contains 3214 entries (2007 release) whilst BindingMOAD contains 9836 protein–ligand structures (4659 different ligands) for which 2964 have binding data (2006 release). In addition, PDBBind includes a "refined set" of 1300 protein–ligand complexes which have been specially selected to provide a set designed for docking and scoring studies.

BindingDB (http://www.bindingdb.org) also has its origins in structure-based design but uses a rather different philosophy when deciding which data to incorporate. Thus whilst PDBBind and BindingMOAD only contain binding data for those protein/ligand systems where a 3-dimensional structure is available, BindingDB does not impose any such limitation. Rather, BindingDB takes as its starting point the collection of biological targets whose structures are available in the PDB or can be accurately modelled. Having selected a protein, the scientific literature is searched (with a focus on the Journal of Medicinal Chemistry and Bioorganic Medicinal Chemistry Letters) to identify relevant data for deposition into the database. In January 2008 BindingDB contained approximately 38,000 measurements for 18,000 molecules and 400 drug targets with the data being extracted from approximately 2200 articles. BindingDB also offers experimental groups the option to deposit results, but so far this has not generated a significant number of entries. Relaxing the requirement that structural data be available for every ligand–protein pair explains the difference in size between BindingDB and the other databases derived from the PDB; in general around 2% of ligands in BindingDB have an exact match in the PDB with 15% of ligands having 90% similarity to a ligand in the PDB.

2.4 Other databases with binding or affinity data

By way of contrast to those databases where structural biology has provided a major impetus there exist a number of sources where binding and affinity data is provided on targets such as G-protein coupled receptors (GPCRs) and ion channels. The GPCR field in particular has benefited from such developments. One widely used source is the GPCRDB [25] (http://www.gpcr.org/7tm) which contains a wealth of information on GPCRs, together with some limited binding data. GLIDA (GPCR-Ligand Database) [26] (http://pharminfo.pharm.kyoto-u.ac.jp/services/glida) is somewhat similar in scope, also containing a limited amount of protein–ligand binding data.

The PDSP Ki database (http://pdsp.med.unc.edu/pdsp.php) currently provides access to approximately 46,000 Ki values for drugs molecules and drug candidates against GPCRs, ion channels, transporters and enzymes. The data comprise those generated within the host laboratory at the University of North Carolina (sponsored by the National Institute of Mental Health) together with data extracted from the literature. The inhibition constant (Ki) values in KiBank [27,28] (http://kibank.iis.u-tokyo.ac.jp) have been extracted from scientific journals (from 1985 onwards) via PubMed searches. KiBank was originally constructed with a structural emphasis and it does include information on 3D protein structure where applicable. However, its collection of more than 16,000 Ki values for more than 5900 chemical structures against more than 100 targets also includes data for proteins without structural information (currently, KiBank covers nuclear receptors, membrane receptors, enzymes, transporters, ion channels and ion pumps). The data in the commercial database BioPrint from Cerep (http://www.cerep.fr) were obtained by systematically profiling approximately 2500 marketed drugs and reference compounds against a large panel of biological assays (including various in vitro ADME end-points). This database thus uniquely provides a full matrix of compounds against biological assays. Applications of BioPrint range from the development of quantitative structure-activity relationships, the identification of associations between *in vitro* and *in vivo* end-points and the use of *in silico* methods to predict potential clinical liabilities.

2.5 Large-scale commercial SAR knowledge bases

In our discussion of commercial knowledge bases we will focus on the following: AurScope databases from Aureus Pharma (http://www.aureus-pharma.com), the ChemBioBase databases from Jubilant Biosys (http://jubilantbiosys.com), the GOSTAR (GVKBio Structure Activity Relationship) database from GVK Biosciences (http://www.gvkbio.com), the KKB (Kinase Knowledgebase) from Eidogen–Sertanty (http://eidogen-sertanty.com), the StARLITe database from BioFocus DPI (http://www.biofocusdpi.com), and the WOMBAT (World of Molecular Bioactivity) [29] database from Sunset Molecular Discovery (http://sunsetmolecular.com). With the initial release in 2001, StARLITe was the first of these databases on the market, followed by the KKB and the AurScope GPCR database in 2002. Jubilant Biosys and GVK Biosciences released their first databases around 2004.

A common approach taken by the providers of commercial knowledge bases is to group together data for a specific group of related proteins or target class (e.g. kinases, proteases, GPCR, etc.). Examples include Eidogen–Sertanty's KKB (Kinase Knowledgebase) or the various databases offered in Jubilant Biosys' ChemBioBase Suite or GVK Biosciences' GOSTAR. Similarly, Aureus Pharma have developed target-class focussed AurScope databases. Most commonly offered are databases for GPCRs, Ion Channels, Kinases, Nuclear Hormone Receptors and Proteases;

data against other target classes such as Transporters, Phosphotases, Lipases, and Phosphodiesterases are only covered by a limited selection of the commercial knowledge bases. However, not all of the knowledge bases in this section are structured in a target class manner; StARLITe and WOMBAT exist as one database where biological activities from various target classes are combined. In addition to the target-class oriented knowledge bases, other databases are also available (e.g. an ADME/DDI AurScope database from AUREUS Pharma, an Antibacterial Database from Jubilant Biosys and a database of pharmacokinetic measurements from WOMBAT).

A key consideration when comparing the databases is the update frequency. Here, there is a trade off between the desire for the database to be as up-to-date as possible and the resources required to install any updates. StARLITe is updated on a monthly basis, whereas GOSTAR, the KKB, the AurScope databases, and the GPCR and Kinase ChemBioBase databases are updated quarterly. WOMBAT is updated twice a year, and the Protease, Ion Channel, and Nuclear Hormone Receptor ChemBioBase databases are updated once a year. Where on-line versions of the knowledge bases are available, the update frequency can be greater since no installation on the part of the customers is required. For example, GVK Biosciences plan to update their online-accessible version of GOSTAR on a fortnightly basis.

With regard to the coverage of published activity data there are major differences in scope across the databases. For example, whereas activity data in StARLITe and WOMBAT is solely derived from journals, AurScope, ChemBioBase, GOSTAR, and KKB are populated from both journals and the patent literature. With minor exceptions, only patents that are filed in English are searched and subsequently indexed if they contain biological activity data. However, some databases also include other structures without any associated biological activity. For example, the ChemBioBase databases capture all molecules that are exemplified in a patent. Eidogen–Sertanty go further as they combine the information about synthetic procedures and reported building blocks to enumerate the chemical space of a given patent. With the exception of Eidogen–Sertanty's KKB database synthesis-related information is only referenced via links to the primary publication, which is done, for example, through links to PubMed (http://www.ncbi.nlm.nih.gov/PubMed) or the use of the DOI (Digital Object Identifier) System (http://www.doi.org). In the KKB database chemical synthesis steps are captured when detailed experimental procedures are reported in the primary references.

Considering the number of scientific journals in the area of Drug Discovery, it is not surprising that only a subset of these journals are comprehensively indexed by database providers. Generally, they identify a set of core journals from which every issue is manually analysed with comprehensive coverage being claimed for a given period. A larger set of journals may also be examined for biological data, but only in an automated fashion (by keyword searches, for example) and so comprehensive coverage cannot be realised. Unsurprisingly, the journals that are considered to be of greatest importance in terms of coverage are the Journal of Medicinal Chemistry (first published in 1959) together with Bioorganic Medicinal Chemistry Letters (first published in 1991) and Bioorganic Medicinal Chemistry (first published in 1993). However, even for these journals complete coverage of all volumes is only rarely claimed (e.g. by GVK Biosciences). Jubilant Biosys' Chem-BioBase databases cover all three journals comprehensively from 1995 onwards and Eidogen–Sertanty's KKB database provides cover of all three journals from 2000 onwards. The practice with regards to journals that are indexed but not necessarily comprehensively covered is again very different. Whereas the content of the ChemBioBase databases is based on biological activity data from the three journals mentioned above, GVK Biosciences databases contain biological activities from more than 2000 journals. Other databases are positioned somewhere between these two, for example StARLITe and WOMBAT with around 10 journals in total and the contents of the AurScope databases being compiled from approximately 350 journals.

The coverage of the published biological activity space described above results in a substantial number of SAR data points: Eidogen–Sertanty's KKB database contains more than 350,000 SAR points (for more than 475,000 molecules), WOM-BAT contains around 417,000 SAR points (for approx. 178,000 unique structures), the StARLITe database contains around 1.7 million SAR points (for approx. 400,000 unique molecules), the AurScope databases contain around 1.9 million SAR points (for approx. 471,000 molecules), the ChemBioBase databases contain 2.8 million SAR points (for approx. 2 million molecules), and GVK Biosciences' GOSTAR contains around 5 million SAR points (for approx. 2.1 million unique molecules).

In addition to coverage it is also important to consider the annotation of the biological activity data. For example, a user might only be interested in molecules whose ion channel activities have been recorded using patch-clamp electrophysiology protocols and that act as antagonists. Not surprisingly, the various providers have chosen different approaches. Whereas Biofocus DPI have decided to include brief descriptions of assay protocols Aureus Pharma and GVK Biosciences capture assay details in approximately 50 individual data fields. This has the advantage of enabling the user to perform quite elaborate searches and also to access experimental details without immediately having to refer to the primary literature.

3. COMPARISON AND INTEGRATION OF DATABASES

Aside from the quantity of data contained within each database, two obvious questions that arise concern the accuracy of the information and the overlap between the various databases. The accuracy is very difficult to quantify but one must assume that such database systems will contain errors and that the primary source must be accessed to obtain the "definitive" data. Nevertheless, the 1% discrepancy that was found to exist when the PDBBind and BindingMOAD databases were compared gives some indication as to the possible error rate for a curated database. With regard to overlap, again this is difficult to quantify, though as many of the major SAR knowledge bases are based upon the same core set of journals, one would expect a significant degree of overlap in terms of content.

One attempt to quantify the overlap and complementarity between a number of public and commercial databases (at least at a structural level) was reported by

Southan and colleagues [30] at AstraZeneca. They compared GVKBIO, WOMBAT, DrugBank, PubChem, MDDR, CMC, BioPrint and a database of natural product molecules (the Dictionary of Natural Products, http://dnp.chemnetbase.com). The numbers of unique structures in each database was determined using a series of in silico filters; this then enabled an overlap matrix to be constructed giving the numbers of compounds common to any pair of databases. Of particular interest was a three-way comparison of the PubChem, GVKBIO and WOMBAT databases. The numbers of unique structures contained within each of these was approximately 7 million, 1.5 million and 130,000 respectively with just over 86,000 structures being common to all three databases. A large number of the structures were unique to PubChem ($\sim$ 6.8 million) as might be expected due to the many structures contributed to that particular database by vendors of compounds for screening. Also of interest was the number of structures unique to GVKBIO (more than 1 million) which were ascribed to compounds extracted from the patent literature.

Consistent with the maxim that "the whole is greater than the sum of the parts" it is often desired to integrate data from a number of sources. A typical drug discovery organisation will have activity data generated by its own in-house screening campaigns. Published SAR data may also be available via one or more of the knowledge bases covered above. The challenge is to combine these data so that all the relevant information can be presented to a user, whilst retaining the integrity of the original. One example of the way in which information from several databases can be used to enhance a corporate collection was described by researchers at Novartis [31] who linked the 2.5 million compounds in the Genomics Institute of the Novartis Research Foundation to databases including PubChem and the World Drug Index, together with various biology and genomic databases. This work enabled potential mechanisms of action to be assigned to compounds and also facilitated the identification of novel chemical scaffolds for drug discovery projects.

A large-scale integration of SAR data was described by Pfizer scientists, who constructed a data warehouse of 4.8 million non-redundant chemical structures together with associated biological data from internal Pfizer sources, the StAR-LITe database, BioPrint and the Current Drugs Investigational Drugs Database (IDDB) [32]. The integration was chemo-centric, in that chemical structures (in the form of SMILES strings) were the key to the storage and retrieval of information. Other important data were protein sequence (i.e. of the target in a bioassay) and disease indication. A critical aspect of the work was the cleaning, mapping and standardisation of the data, commonly referred to as Extraction, Transformation and Loading (ETL). It is at this stage that data fields were selected, data quality issues were addressed, different ways of referring to the same entity were rationalised and controlled vocabularies were introduced. Once complete the data warehouse was used to investigate the relationships between proteins and chemical structures (*vide infra*).

4. APPLICATIONS OF SAR KNOWLEDGE BASES

We shall say relatively little about the use of SAR knowledge bases as a readily accessible source of information about compounds and targets, in part because such use is rarely published and also because it might be considered somewhat "obvious". Nevertheless, the benefits in terms of time and effort saved when undertaking such tasks should not be underestimated. An extension beyond the simple identification and retrieval of individual compounds is the use of such databases as the "feedstock" for any of a variety of *in silico* virtual screening methods in order to create focussed sets of compounds, involving first the identification of relevant active (and possibly inactive) compounds from the knowledge base. These compounds act as query molecules for subsequent 2D or 3D similarity searches or for the construction of some form of mathematical model of activity, all of which can be used to identify compounds from internal or external sources for screening in relevant biological assays.

Among the most widely reported applications of SAR knowledge bases to date has been as an aid in the development and evaluation of in silico modelling methods. In addition to the obvious advantage conferred by not having to construct a large data set from scratch the use of such databases provides a degree of standardisation that enables different methods to be compared. An example is the comparison of 14 different scoring functions using the PDBBind data set reported by Wang and colleagues [14]. By analogy with the curated data sets of protein–ligand complexes that are now widely used for the evaluation of docking algorithms we can anticipate that certain lists of structures together with the associated binding data will emerge as "standard" sets that should be used to evaluate any new scoring algorithm. The challenge, of course, is to avoid over-training such algorithms such that good performance on the standard set cannot be reproduced on unseen data. Other recent examples that make use of the SAR knowledge bases highlighted in this review include the development and evaluation of models of hERG activity based on data from WOMBAT and PubChem [33] and a similar study using data from Aureus Pharma [34], the development of 3D pharmacophore models for histone deacetylase inhibitors based on compounds extracted from the GVK MediChem database and tested against the NCI database [36], the development of recursive partitioning models for CYP1A2 and CYP2D6 inhibition based on data from Aureus Pharma [35], a method for virtual screening that used Self-Organising Maps to achieve novelty that was evaluated against WOMBAT [37], the evaluation of a 3D pharmacophore approach for virtual screening that also used WOMBAT [38], the validation of a naïve Bayes classification method using the Prous Integrity database, WOMBAT and in-house screening data [39] and a comparison of ligand-based virtual screening methods in four different scenarios that used MDDR and WOMBAT [40].

Large-scale integration of chemical and biological data enables a number of interesting chemogenomic analyses to be performed [41,42]. Reduced to its simplest form, such analyses involve the construction of a two-dimensional matrix of biological targets against compounds, with elements of the matrix being the binding constant (or some other measure of biological activity). Invariably, such matrices are sparse, as most compounds will have data for just one or at most a few targets. One notable exception is the BioPrint database which contains percent inhibition values for molecules at a single high (10 μM) concentration. A data set comprising 1567 structurally diverse molecules against 92 assays from BioPrint was used by Fliri and colleagues to derive so-called Biospectra [43] that correspond to a vector of the inhibition values for a given structure. Such descriptors could be used to cluster compounds into families based on patterns of similar biological activity and also to cluster proteins based on patterns of similar interaction with the ligands. It was observed in particular that the biospectra-based clustering separated the compounds into groups that were structurally and pharmacologically related and as such could be used to provide an indication not only of molecular properties but also of the biological response. With such data sets it also becomes feasible to perform "target fishing" [44,45] in which putative targets for a given chemical structure can be identified, in a reversal of the usual goal in drug discovery (i.e. to identify compounds for a given target). The large-scale data integration project described by Paolini and colleagues [32] enabled a number of chemogenomic analyses to be performed, including investigations into the linkages between proteins and protein families, predictions of compound pharmacology, relationships between molecular properties and target class and an estimation of the druggability of a given target.

5. FUTURE PROSPECTS

Whilst Sir Francis Bacon's maxim that "knowledge is power" is invariably applied to situations involving personal or political advancement, it is nevertheless also pertinent to drug discovery, at least insofar as it refers to the extraction of meaningful insights from the available information. However, access to data is not sufficient; it is also important to understand the context, to be wary of over-interpretation and to use appropriate analysis techniques. The development of the knowledge bases described in this review does nevertheless represent an important advance because they relieve scientists of the burden associated with data collection and curation and because the volumes of data they contain enable statistically-significant analyses to be performed. Of course, one has to be careful not to become over-reliant upon such secondary sources; the primary literature should be consulted where appropriate and indeed any data upon which critical decisions are to be made should preferably be reproduced in-house. Another emerging theme is the free availability of large data sets such as PubChem and ChemBank which provide large amounts of data hitherto available only within the commercial pharmaceutical section or from commercial database vendors. Over the next few years we will undoubtedly see further developments, both of the databases themselves and of methods for their integration and analysis.

ACKNOWLEDGMENTS

We would like to thank the following for providing us with information that greatly facilitated the detailed comparison of the main knowledge bases: Mary Donlan (Aureus Pharma), Rosie Hutchings (BioFocus DPI), Sooriya Kumar (Jubilant Biosys), Steven Muskal (Eidogen–Sertanty), Tudor Oprea (Sunset Molecular Discovery), Jagarlapudi Sarma (GVK Biosciences), Stephen Yemm (Symyx).

REFERENCES

1. Scior, T., Bernard, P., Medina-Franco, J.L., Maggiora, G.M. Large compound databases for structure–activity relationships studies in drug discovery. Mini Rev. Med. Chem. 2007, 7, 851–60.
2. Jonsdottir, S.O., Jorgensen, F.S., Brunak, S. Prediction methods and databases within cheminformatics: Emphasis on drugs and drug candidates. Bioinformatics 2005, 21, 2145–60.
3. Olah, M., Oprea, T.I. Bioactivity Databases. In: Triggle, D., Taylor, J., editors. Comprehensive Medicinal Chemistry II, vol. 3. Oxford: Elsevier; 2006, pp. 293–313.
4. Wheeler, D.L., Barrett, T., Benson, D.A., Bryant, S.H., Canese, K., Chetvernin, V., Church, D.M., DiCuccio, M., Edgar, R., et al. Database resources of the National Center for Biotechnology Information. Nucleic Acids Res. 2007, 35, D5–12.
5. Seiler, K.P., George, G.A., Happ, M.P., Bodycombe, N.E., Carrinski, H.A., Norton, S., Brudz, S., Sullivan, J.P., Muhlich, J., Serrano, M., Ferraiolo, P., Tolliday, N.J., Schreiber, S.L., Clemons, P.A. ChemBank: A small-molecule screening and cheminformatics resource database. Nucleic Acids Res. 2007, 36, D351–9.
6. Austin, C.P., Brady, L.S., Insel, T.R., Collins, F.S. NIH molecular libraries initiative. Science 2004, 306, 1138–9.
7. Huryn, D.M., Cosford, N.D.P. The Molecular Libraries Screening Center Network (MLSCN): Identifying chemical probes of biological systems. Ann. Rep. Med. Chem. 2007, 42, 401–16.
8. Tetko, I.V. The WWW as a tool to obtain molecular parameters. Mini Rev. Med. Chem. 2003, 3, 809–20.
9. Brooksbank, C., Cameron, G., Thornton, J. The European Bioinformatics Institute's data resources: Towards systems biology. Nucleic Acids Res. 2005, 33, D46–53.
10. Wishart, D.S., Knox, C., Guo, A.C., Shrivastava, S., Hassanali, M., Stothard, P., Chang, Z., Woolsey, J. DrugBank: A comprehensive resource for in silico drug discovery and exploration. Nucleic Acids Res. 2006, 34, D668–72.
11. Wishart, D.S., Knox, C., Guo, A.C., Cheng, D., Shrivastava, S., Tzur, D., Gautam, B., Hassanali, M. DrugBank: A knowledge base for drugs, drug actions and drug targets. Nucleic Acids Res. 2007, 36, D901–6.
12. Leach, A.R., Shoichet, B.K., Peishoff, C. Prediction of protein–ligand interactions. Docking and scoring: Successes and gaps. J. Med. Chem. 2006, 49, 5851–5.
13. Wang, R., Fang, X., Lu, Y., Wang, S. The PDBBind database: Collection of binding affinities for protein–ligand complexes with known three-dimensional structures. J. Med. Chem. 2004, 47, 2977–80.
14. Wang, R., Lu, Y., Fang, X., Wang, S. An extensive test of 14 scoring functions using the PDBBind refined set of 800 protein–ligand complexes. J. Chem. Inf. Comput. Sci. 2004, 44, 2114–25.
15. Wang, R., Fang, X., Lu, Y., Yang, C., Wang, W. The PDBBind database: Methodologies and updates. J. Med. Chem. 2005, 48, 4111–9.
16. Hu, L., Benson, M.L., Smith, R.D., Lerner, M.G., Carlson, H.A. Binding MOAD (Mother of All Databases). Protein Struct. Func. Bioinf. 2005, 60, 333–40.
17. Smith, R.D., Hu, L., Falkner, J.A., Benson, M.L., Nerothin, J.P., Carlson, H.A. Exploring protein–ligand recognition with binding MOAD. J. Mol. Graph. Model. 2006, 24, 414–25.
18. Chen, X., Liu, M., Gilson, M.K. BindingDB: A web-accessible molecular recognition database. Comb. Chem. High Throughput Screen 2001, 4, 719–25.

19. Chen, X., Lin, Y., Liu, M., Gilson, M.K. The binding database: Data management and interface design. Bioinformatics 2002, 18, 130–9.
20. Chen, X., Liu, M., Gilson, M.K. The binding database: Overview and user's guide. Biopolymers/Nucleic Acid Sci. 2002, 61, 127–41.
21. Liu, T., Lin, Y., Wen, X., Jorissen, R.N., Gilson, M.K. BindingDB: A web-accessible database of experimentally determined protein–ligand binding affinities. Nucleic Acids Res. 2007, 35, D198–201.
22. Puvanendrampillai, D., Mitchell, J.B.O. Protein–Ligand Database (PLD): Additional understanding of the nature and specificity of protein–ligand complexes. Bioinformatics 2003, 19, 1856–7.
23. Roche, O., Kiyama, R., Brooks III, C.L. Ligand–protein database: Linking protein–ligand complex structures to binding data. J. Med. Chem. 2001, 44, 3592–8.
24. Block, P., Sotriffer, C.A., Dramburg, I., Klebe, G. AffinDB: A freely accessible database of affinities for protein–ligand complexes from the PDB. Nucleic Acids Res. 2006, 34, D522–6.
25. Horn, F., Weare, J., Beukers, M.W., Hörsch, S., Bairoch, A., Chen, W., Edvardsen, O., Campagne, F., Vriend, G. GPCRDB: An information system for G protein-coupled receptors. Nucleic Acids Res. 1998, 26, 275–9.
26. Okuno, Y., Yang, J., Taneishi, K., Yabuuchi, H., Tsujimoto, G. GLIDA: GPCR–ligand database for chemical genomic drug discovery. Nucleic Acids Res. 2006, 34, D673–7.
27. Zhanga, J., Aizawa, M., Amaria, S., Iwasawab, Y., Nakanoc, T., Nakata, K. Development of KiBank, a database supporting structure-based drug design. Comput. Biol. Chem. 2004, 28, 401–7.
28. Nakata, K., Amari, S., Nakano, T. Application of KiBank Database. Chem.-Bio Inform. J. 2006, 6, 47–54.
29. Olah, M., Rad, R., Ostopovici, L., Bora, A., Hadaruga, N., Hadaruga, D., Moldovan, R., Fulias, A., Mracec, M., Oprea, T.I.. In: Schreiber, S.L., Kapoor, T.M., Wess, G., editors. Chemical Biology. From Small Molecules to System Biology and Drug Design. Weinheim: Wiley-VCH; 2007, pp. 760–86.
30. Southan, C., Varkonyi, P., Muresan, S. Complementarity between public and commercial databases: New opportunities in medicinal chemistry informatics. Curr. Topics Med. Chem. 2007, 7, 1502–8.
31. Zhou, Y., Zhou, B., Chen, K., Yan, S.F., King, F.J., Jiang, S., Winzeler, A.E. Large-scale annotation of small-molecule libraries using public databases. J. Chem. Inf. Model. 2007, 47, 1386–94.
32. Paolini, G.V., Shapland, R.H.B., van Hoorn, W.P., Mason, J.S., Hopkins, A.L. Global mapping of pharmacological space. Nature Biotech. 2006, 24, 805–15.
33. Li, Q., Jorgensen, F.S., Oprea, T., Brunak, S., Taboureau, O. hERG classification model based on a combination of support vector machine method and GRIND descriptors. Mol. Pharmaceut. 2008, 5, 117–27.
34. Dubus, E., Ijjaali, I., Petitet, F., Michel, A. In silico classification of hERG channel blockers: A knowledge-based strategy. Chem. Med. Chem. 2006, 1, 622–30.
35. Burton, J., Ijjaali, I., Barberan, O., Petitet, F., Vercauteren, D.P., Michel, A. Recursive partitioning for the prediction of cytochromes P450 2D6 and 1A2 inhibition: Importance of the quality of the data set. J. Med. Chem. 2006, 49, 6231–40.
36. Vadivelan, S., Sinha, B.N., Rambabu, G., Boppana, K., Jagarlapundi, S.A.R.P. Pharmacophore modelling and virtual screening studies to design some potential histone deactylase inhibitors as new leads. J. Mol. Graph. Model. 2008, 26, 935–46.
37. Hristozov, D., Oprea, T.I., Gasteiger, J. Ligand-based virtual screening by novelty detection with self-organizing maps. J. Chem. Inf. Model. 2007, 47, 2044–62.
38. Nettles, J.H., Jenkins, J.L., Williams, C., Clark, A.M., Bender, A., Deng, Z., Davies, J.W., Glick, M. Flexible 3D pharmacophores as descriptors of dynamic biological space. J. Mol. Graph. Model. 2007, 26, 622–33.
39. Watson, P. Naïve Bayes classification using 2D pharmacophore feature triplet vectors. J. Chem. Inf. Model. 2008, 48, 166–78.
40. Hristozov, D., Oprea, T.I., Gasteiger, J. Virtual screening applications: A study of ligand-based methods and different structure representations in four different scenarios. J. Comput.-Aided Mol. Des. 2007, 21, 617–40.
41. Rognan, D. Chemogenomic approaches to rational drug design. British J. Pharmacol. 2007, 152, 38–52.
42. Ekins, S., Mestres, J., Testa, B. In silico pharmacology for drug discovery: Methods for virtual ligand screening and profiling. British J. Pharmacol. 2007, 152, 9–20.

43. Fliri, A.F., Loging, W.T., Thadeio, P.F., Volkmann, R.A. Biological spectra analysis: Linking biological activity profiles to molecular structure. Proc. Natl. Acad. Sci. USA 2005, 102, 261–6.

44. Jenkins, J.L., Bender, A., Davies, J.W. In silico target fishing: Predicting biological targets from chemical structure. Drug Discovery Today: Technologies 2006, 3, 413–21.

45. Nettles, J.H., Jenkins, J.L., Bender, A., Deng, Z., Davies, J.W., Glick, M. Bridging chemical and biological space: "Target fishing" using 2D and 3D molecular descriptors. J. Med. Chem. 2006, 49, 6802–10.

CHAPTER **12**

PubChem: Integrated Platform of Small Molecules and Biological Activities

Evan E. Bolton*, Yanli Wang*, Paul A. Thiessen*, and **Stephen H. Bryant*,1**

* National Center for Biotechnology Information, National Library of Medicine, National Institutes of Health, Department of Health and Human Services, 8600 Rockville Pike, Bethesda, MD 20894, USA
1 Corresponding author. E-mail: bryant@ncbi.nlm.nih.gov

Annual Reports in Computational Chemistry, Vol. 4
ISSN 1574-1400, DOI: 10.1016/S1574-1400(08)00012-1

1. INTRODUCTION

PubChem [1], an open repository for experimental data identifying the biological activities of small molecules, is a part of the Molecular Libraries and Imaging (MLI) component of the National Institutes of Health (NIH) Roadmap for Medical Research initiative [2]. This program includes the Molecular Libraries Screening Center Network (MLSCN), grant-supported experimental laboratories, and a shared compound repository, referred to as the Molecular Libraries Small Molecular Repository (MLSMR) offering biomedical researchers access to chemical samples.

PubChem archives the molecular structure and bioassay data from the MLSCN and other contributors. PubChem provides search, retrieval, and data analysis tools to optimize the utility of these results. PubChem further enhances the research utility of the MLSCN output by including other public sources of chemical structure and bioactivity information and by integration of this data with other NIH biomedical knowledgebases. The primary aim of PubChem is to provide a public on-line resource of comprehensive information on the biological activities of small molecules accessible to molecular biologists as well as computational and medicinal chemists.

Initially launched September 2004, PubChem follows the GenBank [3] approach, whereby investigators make direct data submissions. PubChem depends on its contributors to help keep the database as comprehensive, current, and accurate as possible. The processing of PubChem is highly automated, as opposed to being manually curated, keeping the overall database cost low. The open repository nature of PubChem has a 25 year precedent in biology, for example, GenBank, SwissProt [4], PDB [5], etc., but there is less of a precedent for this model in chemistry.

The location of PubChem at the National Center for Biotechnology Information (NCBI) [6] provides the unique ability to integrate directly with a substantial wealth of biomedical information, over thirty databases with information ranging from scientific articles to genes, available within the NCBI Entrez search system [7]. By leveraging and integrating with these resources, PubChem provides a powerful, publicly accessible platform for mining biological information of small molecules.

2. DESCRIPTION

PubChem is organized as three distinct databases: PubChem Substance, PubChem Compound, and PubChem BioAssay. PubChem Substance contains descriptions of chemical samples, provided by data depositors, and links to information on their biological activities. The description includes PubChem Compound identifiers in cases where the chemical structures of compounds in the sample are known. Links providing information on biological activity include those to PubMed [8] citations, protein 3-D structures [9], links to contributor websites, and to biological testing results available in PubChem BioAssay.

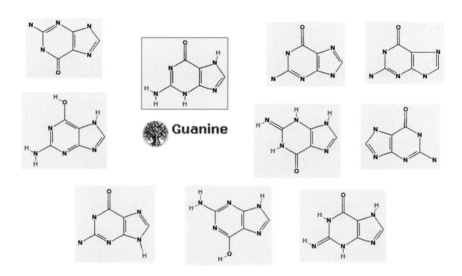

FIGURE 12.1   Different structural representations of guanine deposited in PubChem.

PubChem Compound contains the unique chemical structure content of Pub-Chem Substance. Compounds may be searched by computed chemical properties and are pre-clustered by structure comparison into identity and similarity groups. Whenever possible, compounds are linked via PubChem Substance to information on their biological activities.

PubChem BioAssay contains the results of biological activity testing from a variety of sources. It provides searchable descriptions of each bioassay, including conditions and readouts specific to the screening procedure. PubChem BioAssay provides outcomes for the depositor's tested substances as links to PubChem Substance. Associations between biological testing results and the unique chemical structures are also generated to provide a comprehensive overview of the biological profile of tested compounds.

Abstracting the unique chemical structure content in PubChem Substance to create PubChem Compound is not always trivial. Widely adopted standards or rules for chemical structure representation do not exist, with various groups or individuals adopting preferences based on their organizational needs. Further complicating matters is that PubChem accepts chemical information from a multitude of depositors, each with the potential to represent identical chemical structures in a different way. For example, a molecule as simple as guanine (Figure 12.1) has a number of equivalent representations readily recognizable by a chemist as guanine. Programming a computer to recognize such chemical representations as being the same is nonetheless a challenge.

The normalization method used by PubChem to identify unique chemical content is referred to here as "standardization." This procedure involves a series of automated processing steps, outlined in Figure 12.2, to determine when a provided chemical structure description is well defined and chemically reasonable. The standardization processing steps involve: verification that each atom is a known

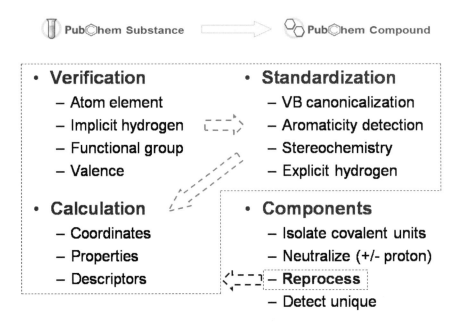

FIGURE 12.2   Overview of the processing performed on chemical structures deposited in PubChem.

element, assignment of implicit hydrogens to organic elements missing valences, normalization of functional group representations, validation that each atom valence and formal charge is reasonable, valence-bond canonicalization for tautomer and resonance invariance, extended aromaticity resonance detection and annotation, stereochemical center identification, and conversion of implicit to explicit hydrogens for unambiguous atom valences. Additional processing is performed to isolate unique covalent units within the chemical sample description of mixtures, which are acid/base neutralized when possible, and reprocessed using the above procedure. Subsequent processing of each standardized structure involves computation of 2-D depiction coordinates and calculation of basic chemical properties (e.g., molecular weight, molecular formula, etc.) and chemical descriptors (e.g., canonical SMILES [10], InChI [11], IUPAC name [12], etc.).

Contributed substance descriptions that do not include a chemical structure or that fail the PubChem chemical structure standardization procedure do not enter or have links to the PubChem Compound database. Prior to analysis or any modification of chemical structure input, care is taken to preserve the original structure description. The result of the normalization methodology employed is a uniform representation of the chemical structure content contained within the PubChem Substance database.

3. DATA RELATIONSHIPS

The fundamental relationships between the three PubChem databases are straightforward. PubChem Substance identifiers (SIDs) relate to PubChem Compound identifiers (CIDs) through chemical structure standardization. Each substance, if it standardizes, will have a corresponding CID that is the main "standardized" form of that substance, representing the whole structure. There may also be "component form" CIDs that include unique covalently bonded units, when the substance is a mixture, or an acid/base charge-neutralized form, when the substance is ionized. A parent compound is assigned to each CID, when possible, to identify the primary organic component. PubChem Assay identifiers (AIDs) contain activity data for SIDs. If a substance is associated with a compound, the assay outcome for the SID can be associated implicitly with a CID, as well.

A critical concept for the advanced PubChem user is that of combining and transforming sets of identifiers between the three PubChem databases, based on the above identifier relationships. For instance, there is a many-to-one relationship between SIDs and "standardized" CID, as more than one Substance depositor may have supplied the chemical structure that standardizes to a given CID. (In fact, even within a particular depositor's records, there may be redundant structures because of different sample origins, tautomeric forms, etc.). Also, the perceptive reader will notice there is not a direct relationship between BioAssay (AID) and Compound (CID) identifiers. To discover assays linked to a CID, there is an expansion of that CID to all SIDs for which that CID is the standardized form; AIDs can be associated with CIDs linked to any of these SIDs.

Many of the PubChem tools perform such transformations of the ID space implicitly, such as assay tools that work with sets of CIDs, or Entrez searches of CID chemical property indices in PubChem Substance, like IUPAC name, that actually come from standardized compounds. It can be important to understand these implicit relationships when navigating through PubChem, especially when searching and analyzing records across multiple databases.

As of March 2008, PubChem contains more than: 1,000 bioassays, 28 million bioassay test outcomes, 40 million substance contributed descriptions, and 19 million unique compound structures contributed from over 70 depositing organizations. While the majority of screening data were contributed by NIH funded screening centers under the MLSCN network, PubChem BioAssay database also contains test outcomes from a number of other organizations, including the sixty tumor cell line assays from DTP/NCI [13], toxicity data from the DSSTox program at EPA [14], and bioactivity data extracted from literature by the BindingDB project [15].

4. INTERFACE

The primary interface to PubChem data is through the NCBI search engine, Entrez. This web-based interface is simple, yet powerful, with many features not immediately apparent to those unfamiliar with the Entrez system. This section is intended as both an introduction and a guide to the more advanced Entrez features, and the types of Entrez PubChem queries that can be performed.

4.1 Entrez

There are a number of entry points to Entrez. The simplest is to go to the NCBI home page (http://www.ncbi.nlm.nih.gov/) from which one can input a search term (or terms) and initiate a search by activating the 'Go' button. By default, if a specific database is not selected in the search menu, the search is performed across all +30 databases available within Entrez, of which PubChem is a part. This "global query" result lists the count of records for the query in each of the Entrez databases. To see the PubChem query results, simply select one of the three Pub-Chem databases (Substance, Compound, or BioAssay), and a detailed report for records matching the query is displayed for that database. One can also begin at the PubChem home page (http://pubchem.ncbi.nlm.nih.gov/) where an equivalent search of one of the three PubChem databases may be initiated through the input form at the top.

Figure 12.3 shows the result of searching for the word "aspirin" in Entrez's PubChem Compound database. This default display of multiple records in Entrez is referred to as a document summary (DocSum) report and is common to all Entrez databases. At the top are the common Entrez controls (database selection and search input box) and tabs for other Entrez tools (e.g., Limits, History, etc.) some of which are described in more detail below. Note that the format of this page evolves over time, but the basic controls remain the same. Moving down the Doc-Sum page, the next section contains controls to change the display type; the default is "Summary" (as shown). Each Entrez database has report styles that vary in type and detail of information shown, the overall format is the same—a list of records,

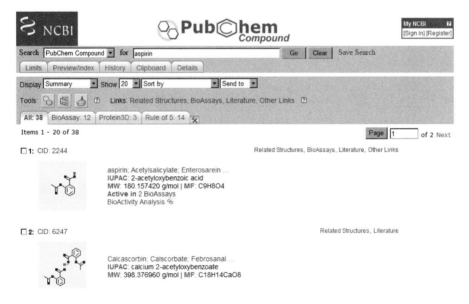

FIGURE 12.3   Partial view of an Entrez document summary (DocSum) report page for the PubChem Compound query "aspirin".

each with report-specific information displayed. Also, controls exist to enable one to sort the results by various means or to export the DocSum to a file or printer.

PubChem databases have a number of additional controls that operate on a query result list, such as icon buttons (provided after "Tools") for assay data analysis and chemical structure download. There are pop-up link menus (provided after "Links" on the same line as "Tools") that provide powerful query result list operations. Also, the pop-up link menus exist for each record, but they function only on the individual record. The meaning of these links is detailed further in the following sections.

The last set of tabs shown in Figure 12.3 (e.g., All, BioAssay, Protein3D, etc.) before the actual record summaries are filters that apply to the current result list. For example, in Figure 12.3, the "BioAssay: 12" tab indicates 12 of the 38 results have associated bioassay data. This tab, if selected, will indicate which 12 records have associated BioAssay data and will allow the result list to be refocused to consider only the 12 records by clicking the "push pin" icon on the tab that appears. These filter tabs are in fact customizable through the "MyNCBI" system. One can create and store new filters in MyNCBI that can be applied to any search in a given database. As depicted in Figure 12.3, the MyNCBI tool is accessed via the box in the upper right corner of the page.

The remainder of the Entrez DocSum page contains the paginated list of results of the current search. While the details will vary according to the specific database and report style, information is shown for each record matching the query result. In the case of the example in Figure 12.3, this includes links to the detailed summary page of each item and other Entrez database records associated with those in the current database.

4.2 Advanced features

Entrez is basically a multi-database search engine. Under the surface are a vast number of details on which fields are available to be searched, what the many types of links mean, how the core Entrez controls function, and so on. All of this may be a bit daunting to the casual user, but understanding these details unlocks the true power of Entrez. This section serves as a guide to PubChem's Entrez databases, including what indices, links, and filters are available, and how these combine together to create an advanced query refinement system.

4.2.1 Entrez indices

An index is a piece of information tied to individual records and matched directly to a user's query in an Entrez search. Each index consists of text, numeric, or date values. Each Entrez database has its own set of indices. These indices are named according to the type of information they contain, for example, the indices "IUPACName" or "MolecularWeight" in PubChem Substance and Compound. Some indices may have multiple values for each record. For example, the index "Synonym" corresponds to chemical or common names of a substance, any number of which may be supplied by the depositor.

By default, when one enters a simple query in the Entrez search interface, that query is matched against all indices in that database. For example, if one searches "aspirin" in PubChem Compound, Entrez will report back any records with an index that contain "aspirin" as (any word in) a synonym, a depositor comment, etc. This is why a text search for "aspirin" also currently brings up the structure of acetaminophen, considering one of the names supplied by a depositor for acetaminophen is "Aspirin-Free Anacin," and so an unrestricted search for "aspirin" will match this record, as well.

It is possible to narrow the search to a particular index by adding the index name in brackets after the term itself. For example, "aspirin[CompleteSynonym]" returns only a single record, the actual structure of aspirin, because only that record has a synonym that matches that query exactly. Also, this shows that some Entrez indices are configured to require an exact match to the entire index, while others allow matches to any individual word in the longer text.

For numeric indices, one can perform a search for a range of values by using minimum and maximum values separated by a colon and followed by the index name in brackets. For example, to find all chemical structures in PubChem Compound with a count of hydrogen bond donors between 0 and 5, the range query would be "0:5[HydrogenBondDonorCount]." In the case of floating point range queries, such as finding all chemical structures with a molecular weight between 214.31456 and 215.0 g/mol, one would use the query "214.31456:215[Molecular-Weight]."

Multiple indices may be searched simultaneously using Entrez's Boolean operators. For example, a query in PubChem Compound of "Br[Element] AND 1[CovalentUnitCount]" will find all chemical structures containing the element bromine and that are not part of a mixture. Please note that Entrez Boolean operators are capitalized (e.g., "AND," "OR," and "NOT").

By default, Entrez removes whitespace, some punctuation, and other special characters from the query string. To make sure Entrez treats the query as a single word or phrase, despite special characters, simply enclose the query in quotation marks. For example, to search the PubChem Compound database using the InChI string of aspirin, one would use ""InChI=1/C9H8O4/c1-6(10)13-8-5-3-2-4-7(8)9(11)12/h2-5H,1H3,(H,11,12)/f/h11H"[InChI]" as the query.

Knowing what indices are available in a database is the key to maximizing the power of an Entrez search. The indices may be listed by going to the "Preview/Index" tab in Entrez, and opening the menu on the bottom left. Also, this page provides an interface for constructing index-specific queries. A complete list and description of the Entrez indices available for the three PubChem databases are detailed in the "Indices and Filters in Entrez" section of the help documentation: (http://pubchem.ncbi.nlm.nih.gov/help.html).

4.2.2 Entrez links
In addition to index searching, Entrez provides cross links or associations between records in different Entrez databases, or within the same database. These links may be applied to an entire search result list, via the links pop-up menus at the top of

a DocSum page (see Figure 12.3), or to an individual record, via link menus on the right side of each entry in the DocSum.

Links provide a way to discover relevant information in other Entrez databases based on a user's specific interests. Equivalently, one may think of this as a way to transform an identifier list from one database to another based on a particular criteria. From PubChem Substance, for example, the link "PubChem BioAssays, Active" provides all assays where that particular substance (or any substance within a multi-record list) was found to be active, where the meaning of "active" is specific to and defined by a particular assay depositor. In a similar cross-database fashion, activating the "PubChem Same Substances" link from a PubChem Compound record will lead to all deposited substances exactly matching that compound, providing a method to see which depositors deposited a particular compound. Some links operate within the same database, going to records that are related in some way. For example, the "Similar Compounds" link from a structure in PubChem Compound will take the user to a DocSum display of all compounds that have a 2-D Tanimoto-based similarity of at least 90% to the structure.

4.2.3 Entrez filters

Filters are essentially Boolean bits (true or false) for all records in a database that indicate whether or not a given record has a particular property. Filters may be used to subset other Entrez searches according to this property, by adding the filter to the query string. For example, the "pcsubstance_pcassay" inter-database filter has a "true" bit for every substance that has associated PubChem BioAssay data, such that a search for "100:200[MolecularWeight] AND pcsubstance_pcassay[Filter]" in PubChem Substance will return a list of all substances with molecular weight from 100.0 to 200.0 g/mol and that have associated PubChem BioAssay data.

Filters are related to links in that the majority of filters in the PubChem databases are generated automatically based on the presence of links. In the above example the "pcsubstance_pcassay" filter has a "true" bit for every substance for which a PubChem BioAssay link is present (e.g., in the pop-up menus of the Entrez DocSum for that substance).

There are some special filters that are not link-based. The query "all[Filter]" simply returns every record in a given Entrez database. A database may have other special filters defined, such as the "has_pharm" filter in PubChem Compound that indicates whether a given chemical structure has a known pharmacological action.

The filters for each Entrez database may be listed by going to the "Preview/Index" tab in Entrez, opening the menu on the bottom left, selecting "Filters," and pressing the "Index" button. Also, this page provides an interface for adding filters to Entrez queries. A complete list and description of the custom Entrez filters available for the three PubChem databases are detailed in the "Indices and Filters in Entrez" section of the help documentation (http://pubchem.ncbi.nlm.nih.gov/help.html).

4.2.4 Entrez history

Entrez is a query refinement engine. In addition to enabling complex searches across databases, as described above, Entrez has a history mechanism (Entrez history) that automatically keeps track of a user's searches, temporarily caches them (for eight hours), and allows one to combine search result sets with Boolean logic. For example, say a structure search (described elsewhere in this document) has been completed, resulting in a list of 10,000 compounds. One may wish to narrow this search by other means, such as to find all compounds in the original search result that satisfy the "Lipinski Rule of 5" [16]. To do this, one would go to the "History" tab in Entrez, where all recent searches are listed, and find the history number in the leftmost column corresponding to the structure search in question (e.g., something looking like "#5 : 10,000 document(s)"). Then, in the search form, at the top of the page, one would use this history number to formulate a query such as "#5 AND lipinski_rule_of_5[Filter]," to narrow the original result to only those records that satisfy both the original query and the "Lipinski Rule of 5."

Entrez history is used heavily by PubChem tools (which are not a part of Entrez) so results of user searches can be used as a subset for further manipulation. For example, the chemical structure download service (described below) reads Entrez history items, so one can generate an SDF file containing just those compounds found in a PubChem Compound Entrez result set. For example, the BioAssay tools (also described below) make frequent use of Entrez history, so that structure queries can be used to subset assay results in a chemical structure analog series.

It is important to note that Entrez history is database-specific. One cannot use it to combine search results between databases (e.g., to 'AND' together a CID list with an AID list). Cross-database links must first be used as set transformation operators, so all ID lists are in the same database. For example, following the "Pub-Chem BioAssays" link from a set of CIDs will create a new set of AIDs that have any test results for the set of CIDs (again with the implicit understanding that CID is first expanded to SID, which is built into the CID-AID links). From there, one may combine this set of AIDs with other search results in the BioAssay database using the Entrez Boolean logic.

Understanding which ID space transformations are implicit and which may be performed explicitly through links or other tools, is crucial to successful use of the advanced PubChem tools. With Entrez history, the user has complete control over the set logic used in sophisticated query refinement. Both of these concepts become even more important when dealing with the PubChem programmatic tools (described below).

5. TOOLS

We have described how PubChem databases are integrated into Entrez, enabling detailed and flexible searches across the PubChem data; however, Entrez is essentially a text search engine and is not amenable to more detailed chemical and bioassay data analysis. Such analysis must be handled by specialized applications.

As the PubChem data content grows, there is an ever increasing need for facile methods of efficient large-scale data management and analysis.

Researchers require the ability to obtain comprehensive summaries of the biological activities of small molecules. In addition, scientists are interested in other chemicals which share structural or physical property similarities to known bioactive entities, or have similar biological activity profiles. To this end, the PubChem BioAssay system provides additional data analysis tools for utilizing and analyzing the biological activity data. These include tools for comparison of test results across multiple experiments, visualizing and exploring structure-activity relationships, and summarizing bioactivity information.

There are two general categories of specialized applications provided by PubChem: those that deal with chemical structure information and those that deal with bioassay data. These categories are not totally distinct; however, as several of the PubChem tools, such as structure-activity analysis and structure clustering, directly bridge the two. These particular tools are closely integrated with Entrez, as searches in one may be used as starting points in the other, but they are conceptually and operationally separate from Entrez. The goal of this section is to describe available tools and how they combine together to form a unified platform for mining PubChem chemical and biological data.

5.1 Summary pages

The Entrez DocSum reports serve a limited quantity of data to help navigate and subset records. Detailed information is provided by PubChem summary pages. Each record in an Entrez DocSum contains a link that leads to the more detailed information on a specific record. Typically these pages are reached through Entrez, but one can also navigate to them directly. For PubChem Substance SIDs, the summary page URL is of the form:

http://pubchem.ncbi.nlm.nih.gov/summary/summary.cgi?sid=1234

where the SID (substance identifier) is provided as an argument. Similarly, for a PubChem Compound the URL is like:

http://pubchem.ncbi.nlm.nih.gov/summary/summary.cgi?cid=2244.

For a PubChem BioAssay summary page, the URL has the form:

http://pubchem.ncbi.nlm.nih.gov/assay/assay.cgi?aid=910.

In general, summary pages contain the detailed information necessary to understand how individual PubChem records combine information into a comprehensive system of interconnected data.

5.1.1 Compound/substance summary
The layout of the substance and compound summary pages are very similar. The content of this summary is heavily dependent on the information provided by depositors and our ability to integrate contributed information with biomedical

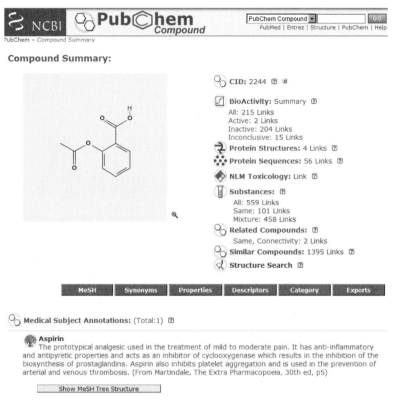

FIGURE 12.4   Partial view of a compound summary for aspirin (CID 2244).

resources at NCBI. Summary pages, despite being continually refined as content is added or usability improved, provide an overall summary of what is known about a particular substance or compound. In general, a compound or substance summary will contain these basic aspects: a depiction of a chemical structure; indication of where or how the record originated (e.g., who contributed the record); links to a set of related inter-database Entrez resources, such as a protein 3-D structures or literature articles; links to known biomedical information (e.g., pharmacological actions of a drug); a list of synonyms or names associated with the record; computed chemical structure properties and descriptors; and record download controls. Figure 12.4 depicts an example of a compound summary for aspirin.

Substance summary pages are distinctly different from compound summary pages in two important ways. First, a substance summary provides access to the depositor's original structure information as well as the standardized form of the substance (when applicable), with the standardized form always shown by default. Second, a substance summary only provides information provided by a single depositor, whereas a compound summary page aggregates information across all depositors providing substances that standardize to that compound.

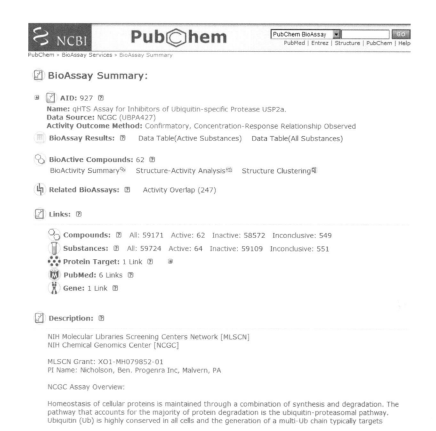

FIGURE 12.5   Partial view of a bioassay summary for a confirmatory (secondary screen) assay for ubiquitin-specific protease USP2a (AID 927).

5.1.2 BioAssay summary

A BioAssay summary displays descriptive information and a summary of the assay results. This includes an overview and background of what the assay attempts to achieve, the assay protocol utilized, references, definition of all reported assay outcomes, indication of the primary result fields, and explanation of the criteria used when considering samples as active or inactive. One can use the "Related BioAssay, Depositor" link to find additional screening performed for a particular assay project. An example bioassay summary is depicted in Figure 12.5.

5.2 Structure search

The PubChem structure search tool enables one to query and subset PubChem Compound by a variety of chemical structure search types and optional filters. The chemical structure search service may be directly accessed using the URL:

http://pubchem.ncbi.nlm.nih.gov/search/.

The supported query input formats for the structure search tool are SMILES, SMARTS [17], InChI, CID (PubChem Compound identifier), molecular formula, and SDF [18]. There is also an online JavaScript-based chemical structure sketcher through which a query may be manually drawn, edited, or imported. The sketcher is compatible with modern web browsers and does not require special software to be downloaded or installed.

Multiple chemical structure search types are available. Identity search enables one to find identical PubChem records at different levels of "sameness" through consideration of structural connectivity and either the presence or absence of isotopic and stereochemical information. Similarity searches locate chemical structures similar to a query, using a percent similarity measure employing the Tanimoto equation [19] and a dictionary-based fingerprint, analogous to the MACSS structure-based keys [20], that are described on the PubChem FTP site: (ftp://ftp.ncbi.nlm.nih.gov/pubchem/specifications/pubchem_fingerprints.txt). Molecular formula searches employ a flexible query containing the count of particular elements in a chemical structure. Substructure searches locate records that contain all atoms in a particular chemical structure query pattern. Superstructure searches locate records that comprise a subset of atoms in a particular chemical structure query pattern.

While the input query and search type are all that are necessary to perform a structure search in PubChem, there are numerous choices by which one may narrow the search to smaller subsets of PubChem. For example, one may search only within a previous Entrez search result, or even a previous structure search result, or upload a file of CIDs against which the search is to be performed. One may filter based on a wide variety of properties, such as molecular weight, heavy atom count, presence or absence of stereochemistry, assay activity, elemental composition, depositor name or category, etc. Most of these subset operations could be accomplished through appropriate Entrez index queries followed by Boolean operations on structure search results; however, the structure search tool provides a convenient one-step interface for chemical search refinement.

All compound structure searches are queued on a set of devoted NCBI servers. The user is taken to a search status page after submitting a query, with a meter showing the relative progress of the requested task. By default, a structure-based query is allowed to take as much time as necessary to complete, but may be limited in the total count of result structures; however, query time and result limits are customizable. Another key feature is the ability to import and export PubChem structure queries to an XML file, which allows one to repeat a particular compound query without filling out the search form again, to share a complex query with a colleague, or to serve as an example for constructing queries for the PubChem Power User Gateway (PUG) interface (described later).

5.3 Structure standardization

Given that PubChem modifies chemical structure information to normalize its representation, it is important for contributors and users of PubChem to explore or understand these changes (e.g., when attempting to integrate external resources with PubChem). With this aim in mind and in the spirit of structure modification transparency, the PubChem chemical structure standardization tool was created. Chemical structure standardization may be directly accessed using the URL:

http://pubchem.ncbi.nlm.nih.gov/standardize/.

This service takes as input a chemical structure and (if standardization is possible) outputs a chemical structure. Allowed structural input and output formats include SMILES, InChI, or SDF file; however, the input and output formats need not be the same. As with structure search, the standardization service is queued on PubChem servers, meaning a request may not start right away or may not complete immediately. One may also import and export standardization requests to a local XML file to serve as an example for constructing queries for the PUG interface (described in detail later).

5.4 Structure downloads

After working with PubChem to achieve a particular subset for a query of interest, it is often important for a user to export resulting substance or compound records from PubChem for further local analysis. The structure download tool prepares PubChem Substance or PubChem Compound records as an export from Entrez in a number of formats. While all PubChem data is available on the PubChem FTP site (via the URL ftp://ftp.ncbi.nlm.nih.gov/pubchem/), being able to interact with a user-selected subset is substantially more convenient. The structure download tool may be directly accessed using the URL:

http://pubchem.ncbi.nlm.nih.gov/pc_fetch/.

Using the download service is straightforward. The user need only perform a search using any combination of Entrez and PubChem-specific search tools, then go to the download page from the PubChem Substance or Compound Entrez Doc-Sum using the download link as indicated by a button with a disk icon. After the user selects an export format, a file containing the exported records will be prepared (on queued PubChem servers, meaning a download may not start or finish immediately) and then served to the user as an URL specifying the download location. It is important to understand that records retrieved from PubChem Substance contain the original deposited information, whereas those from PubChem Compound are standardized forms of the deposited structural information.

A number of formats are available for data export. These formats include SDF, image, small image, SMILES, InChI, XML, and either text or binary ASN.1. The PubChem native archive data format is ASN.1; all other formats are converted from the original ASN.1. The XML formatted data is exactly equivalent to the ASN.1 in content. SDF format is the industry standard for conveyance of chemical structure information and is readily imported into a large number of chemistry programs. Unfortunately, the SDF format is unable to handle all aspects of the ASN.1 data and may not contain all archived information. The PubChem ASN.1 specification, XML schema, and a description of PubChem SDF structure data (SD) tags are all found on the PubChem FTP site in the "specifications" directory.

ASN.1 is a binary format. NCBI utilizes a textual description of ASN.1 that is both computer and human readable (to some extent), but that is not a standard type of ASN.1 data format. This means ASN.1 parsing libraries other than NCBI's may be unable to read it. The PubChem ASN.1 text format does provide a relatively facile means for users to find pertinent information stored in the archive format by simple inspection.

The PubChem download service exports chemical structure images. The images are either 300 × 300 or 100 × 100 pixels in size. The image format is PNG and images are packaged as SID or CID-numbered files in a zip (.zip) archive.

Exports of structural descriptors, SMILES and InChI, provide chemical structure information in a simple tab-delimited text file containing CID or SID and either the isomeric SMILES or InChI strings. Given the very nature of the formats of SMILES and InChI, not all chemical structure information can be identically represented. For example, SMILES encodes only covalent bonds, while PubChem supports the additional concepts of ionic, complex, and dative bonds. Most small molecules in PubChem can be reproducibly interconverted between InChI, SMILES, and PubChem ASN.1 formats without loss of chemical structure information.

Files may optionally be compressed in standard gzip (.gz) or bzip2 (.bz2) formats. Downloads through the structure download tool are limited to a maximum of 250,000 records per request. Image downloads are limited to 50,000 per request due to the inherent limitations of the zip (.zip) format. As with the other structure tools, the structure download service is accessible using the PubChem Power User Gateway (PUG).

5.5 BioActivity analysis

Beyond a summary description, one would like to view, analyze, and display the actual bioassay data. PubChem provides an integrated suite of tools, each presented as an individual tab, for this purpose. One would use the bioactivity summary tool to, at a glance, be able to examine an overview of the bioassays tested for a list of substances or compounds. To be able to subset and analyze substances or compounds tested in a set of bioassays, one would use the structure-activity analysis tool. To view the actual bioassay outcomes, one would use the data table tool.

5.5.1 BioActivity summary

The BioActivity summary tool is a powerful data analysis tool that provides a comprehensive view of biological activity information available for one or more small molecules. It allows one to compare and examine biological outcome counts across multiple assays, enabling common groups of compounds tested in different assays to be rapidly located (e.g., for Structure–Activity Relationship (SAR) analysis). Furthermore, it allows one to select specific test results to view via the 'Data Table' tab and to perform exploratory data analysis via the 'Structure–Activity' tab. Figure 12.6 depicts an example bioactivity summary.

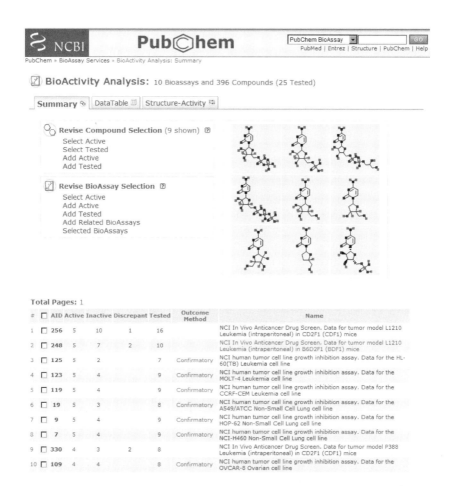

FIGURE 12.6   Partial view of a BioActivity Summary for cytidine (CID 596) and its 2-D similarity neighbors for all bioassays tested within PubChem.

BioActivity summary provides a set of functions that allows one to revise the substance/compound and assay sets. For example, one may focus only on a subset of compounds that are active in one or more of the selected assays using the 'Compound I Select Active' link, or explore additional screen sets where the given compounds were considered active using the 'Assay I Add Active' link. PubChem provides multiple access points for this service. For compounds or substances tested found in Entrez, one can launch this service for each individual record using the direct "BioActivity Analysis" link, or, for all of the records from an Entrez search, through the launching point at the "Tool" area.

5.5.2 Structure–Activity Analysis

Structure–Activity Analysis is an exploratory tool which performs single linkage clustering analysis for small molecules and their biological screening information in a "heatmap" style display. With this web based tool, a list of assays may be provided and clustered based on activity profile of tested compounds or based on protein target sequence similarity. A set of compounds entered can be clustered either by activity spectrum or 2-D chemical structure similarity. Facilities are provided for navigating between various PubChem web tools and Entrez, and can be accessed throughout the heatmap display. For example, one may identify a compound cluster, click the blue circle near the node of a compound cluster, and select "Compound in Entrez" from the pop-up menu to send the compounds back to Entrez for display. One may also "zoom in" on a sub-region in the heatmap display and request test results generated by multiple screenings for a cluster of compounds, using the embedded tool menu. The service provides various "Revise" functions allowing one to change the selection of compounds or assays. Using the "Revise" function, one can continue the analysis by combining additional screening results. With these versatile functions, the Structure–Activity analysis tool provides a powerful service for iterative analysis of the complex screening data and associated chemical and biological information in PubChem and NCBI resources. An example structure activity analysis is provided in Figure 12.7.

The structure clustering aspect of the Structure–Activity Analysis tool is also available as a separate standalone service for the examination of the similarity of a list CIDs. The tool is called Structure Clustering. Considering the functionality is a subset of the Structure–Activity Analysis tool, it is not described in further detail.

5.5.3 Data Table
To obtain the actual test results in a tabular format, with a single compound or substance per row, one uses the Data Table tool. Able to handle multiple assays and multiple compounds or substances, the Data Table provides various means to "collapse" the data view, including compound (as opposed to substance) specific operations, ignoring or including stereochemistry, and grouping by parent compound (for compound salt form invariance). One may also choose how to handle duplicate and conflicting outcomes resulting from the various methods. Pagination and per column sorting controls are available and all data may be exported in different ways.

The Data Table tool is multifunctional with separate tabs for views of concise (only primary results) or all data. Additional controls for plotting bioassay data columns and for subsetting the displayed data using particular data values or data ranges are provided by the "Plot" and "Select" tabs, respectively. Figure 12.8 depicts an example data table view.

6. PROGRAMMATIC TOOLS

While giving access to all available PubChem data and functionality, interactive web-based interfaces are not particularly well suited to highly repetitive or automated tasks. Without programmatic tools, tasks such as performing specific data lookups for a large number of chemical structures would be tedious if not impossible to perform and a software tool that integrates with PubChem services and data would be difficult to create and maintain. With programmatic access to PubChem,

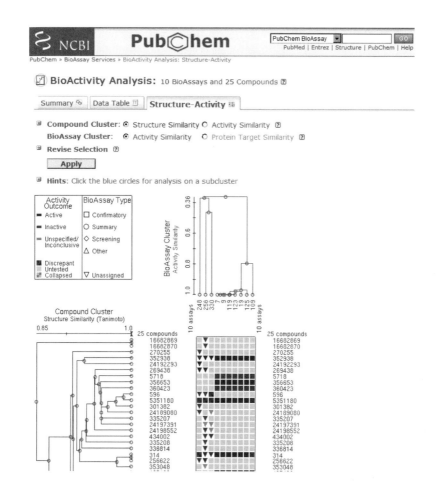

FIGURE 12.7   Partial view of a Structure–Activity Analysis for cytidine (CID 596) and it 2-D similarity neighbors for all bioassays tested within PubChem, with compounds clustered by 2-D structure similarity and assays clustered based on compound biological response.

data can be utilized in more imaginative and complex ways without the need to download the entire PubChem content or to duplicate PubChem functionality.

Two sets of synergistic tools are available for programmatic access to PubChem data, Entrez utilities [21] (eUtils) and the PubChem Power User Gateway (PUG). Entrez-based access is achieved through the use of eUtils. To provide access to the capabilities of PubChem tools, PUG is available. Together, these two facilities enable users to interact with PubChem using XML over HTTP.

6.1 Entrez utilities

Entrez has a suite of associated tools, collectively called eUtils. Together, these tools provide access to nearly all Entrez functionality, primarily through an XML

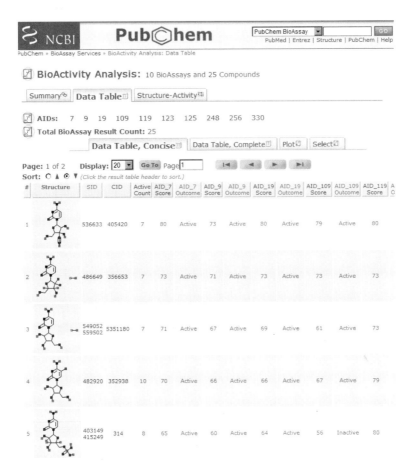

FIGURE 12.8   Partial view of a concise Data Table for cytidine (CID 596) and it 2-D similarity neighbors for all bioassays tested within PubChem.

over HTTP interface. These tools are described in detail elsewhere:

http://eutils.ncbi.nlm.nih.gov/entrez/query/static/eutils_help.html.

The primary eUtil tools of most interest to PubChem users are eSearch, eFetch, ePost, eLink, eHistory, and eInfo. eSearch performs an Entrez search, with the same query syntax as web-based Entrez queries (e.g., to query PubChem Compound for the chemical name "aspirin"). eFetch returns an ID list from a prior search (e.g., the list of PubChem Compound identifiers (CIDs) from the aforementioned query of "aspirin"). ePost creates a new ID list by upload of a list of identifiers (e.g., substance identifiers (SIDs)). eLink follows a given link type to create a new ID list from an existing one (e.g., to find all PubChem BioAssay identifiers (AIDs) associated with a list of SIDs). eHistory returns information on current Entrez History entries. eInfo lists available Entrez indices and links for a given database.

Each of these eUtils applications can return data in XML format for automatic parsing by a script or application. Most eUtil tools have the option to use an Entrez history key, which include a web-environment (WebEnv) and Entrez history item (query_key) as arguments, as input or output. This enables Entrez history to store sets of identifiers temporarily, relieving the user's application of the burden of continually sending and receiving potentially large ID lists. The XML specification, in DTD form, for each eUtil tool can be found at the URL:

http://eutils.ncbi.nlm.nih.gov/entrez/query/DTD/index.html.

6.2 PUG

PubChem's Power User Gateway (PUG) is a single entry point to a vast array of PubChem functionality. It is not necessarily intended for the casual user, but rather for those who are seeking a low-level interface access to PubChem. Outlined here, PUG is documented with examples at:

http://pubchem.ncbi.nlm.nih.gov/pug/pughelp.html.

The basic design of PUG is simple, a central gateway to multiple PubChem functions. PUG does not take URL arguments. All communication with PUG is through XML over HTTP. To perform any request, one formulates input in XML and then sends it to PUG via an HTTP POST. PUG interprets the incoming request, initiates the appropriate action, and then returns results in XML format. With this design, PUG may be used with any scripting or programming language that has the ability to read and write XML, and to send and receive data via HTTP. The XML specification for the XML used by PUG may be found, in DTD and XML schema forms, respectively, at:

http://pubchem.ncbi.nlm.nih.gov/pug/pug.dtd,

http://pubchem.ncbi.nlm.nih.gov/pug/pug.xsd.

Either of these specifications may be used to guide the creation of valid input XML to send to PUG and to parse the returned results. Because PUG encompasses a wide variety of functions, its XML structure is necessarily complex. It may be easier to create input XML data with the help of a tool that can generate program code from or at least validate XML using a DTD or schema.

PubChem tools for structure search, standardization, and downloads are enabled via PUG, with more to be added. In each case, the options available through PUG are the same as those available through the interactive web pages, including all the advanced options and filters of the structure search service. In fact, most of the web tools can write out queries in PUG's XML format, which can be sent directly to PUG or used as templates for constructing new PUG requests.

As with the web-based tools, requests through PUG may be queued on PubChem servers. Thus, PUG may not deliver an answer directly in response to the initial request. Rather, for cases where execution may take some time, PUG will return a waiting message, along with a request identifier which is used to poll

PUG periodically for the status of that request. PUG responds with another waiting message if the operation is still in progress, an error message if it failed, or a success message with the final results, when the task is finished. It is up to the PUG user to add a periodic status check loop to handle these queued requests properly.

The combination of PUG and Entrez eUtils opens up a wide spectrum of programmatic tasks that can harness the true power of PubChem inside custom applications. An advantage to this approach, compared to having a local copy of PubChem data on the user's computer, is that the mass of PubChem data and complexity of the analysis functions are all maintained by PubChem, thus, the CPU cycles needed to perform the tasks are hosted by PubChem. The user needs only this basic interface to access PubChem infrastructure, at the relatively small investment of a little programming.

7. DEPOSITION SYSTEM

PubChem is an open repository. Organizations may contribute information about small molecules and integrate their public resource with PubChem, in part by providing URLs back to and from their website to PubChem. The types of PubChem depositors are greatly varied with contributors from government organizations, academic groups, chemical reagent and screening library suppliers, scientific journals, scientific data publishers, physical property databases, and more. To handle this quantity and diversity of data, PubChem created an on-line data deposition system for rapid contribution of substance and bioassay data. This system may be accessed via the URL:

http://pubchem.ncbi.nlm.nih.gov/deposit/.

Any organization may become a PubChem contributor. The deposition system allows potential depositors to obtain a test account quickly, to examine how their data will look in PubChem and to gain familiarity with the user interface. A test account is nearly identical to a deposition account except data cannot be added to PubChem when using a test account. To actually put data into PubChem, potential depositors must apply for a deposition account. Deposition accounts require a click-thru data transfer agreement that must be agreed upon prior to allowing data to be contributed. Essentially, this agreement enables the depositor to retain all rights to their information while allowing PubChem to display and distribute any provided information.

Deposition of substance information is performed using the industry standard SDF format, which may include using the SMILES or InChI formats as the chemical structure. Depositing properly formatted substance data into PubChem is as simple as uploading a file, via HTTP or FTP.

Deposition of assay information is performed in two parts. Creation of a new assay involves providing a description, protocol, target, readouts, and other associated information using a web-form or via an XML file. After the assay description is completed, assay test results can be readily provided by using the standard CSV (i.e., comma delimited) file format. Assays provide outcomes for substances.

As such, PubChem requires these substances to be available in PubChem prior to providing respective assay information.

Once data is put into PubChem, depositors may update their information at any time. Updates to existing PubChem records cause versioning to occur. Pub-Chem is archival, in that retention of previous versions of records allows PubChem users to access a particular version of a substance or bioassay record, regardless of its revision history. It should be noted; however, that older version information is not presented by default.

Bioassays have two levels of versioning, being major and minor updates. Minor bioassay versions indicate changes to the bioassay textual description. Major bioassay versions indicate addition or reduction in the count of readouts. Major bioassay versioning requires all bioassay data to be completely restated by the depositor, considering the readouts changed in some way. Bioassay records also have substance-level outcome versioning. If a bioassay substance outcome is provided more than once by a depositor, previous reported results are versioned.

8. FUTURE DIRECTIONS

Expansion and enrichment of the bioassay data are ongoing, by adding annotations for small molecules and drugs using publicly available information, such as that provided at the National Library of Medicine (NLM) or the Food and Drug Administration (FDA). With efforts from the scientific community, bioassay data is becoming better annotated by linking target to protein classification resources or molecular pathway information. With further integration with NCBI resources such as PubMed and the Entrez search system, information contained within PubChem will become more discoverable and useful to a broader audience of scientists worldwide.

PubChem currently provides 2D-based data analysis and clustering tools. Small molecules are not flat. They have a rich diversity of 3-D shapes and 3-D orientation of features possible. Addition of a theoretical 3-D description of the PubChem Compound database may open new avenues in the understanding of bioassay outcomes by allowing combination of 2-D and 3-D data analysis and clustering techniques, thus enabling improved hypothesis generation and trend recognition implicit with a biological dataset. Neighboring 3-D descriptions of PubChem Compound, much like the 2-D similarity neighbors currently available, may help scientists identify and better understand interrelationships of the biological properties of small molecules. It is, to this end, that a 3-D description of the PubChem Compound database is in progress.

Programmatic access to PubChem using work-flow automation software (such as Taverna [22] and Pipeline Pilot [23]) and scripting languages (such as Python [24], Ruby [25] and PERL [26]), may enable researchers to make exciting new discoveries and to further leverage and integrate PubChem into their basic research. New interfaces using SOAP-based web services (via WSDL [27]) are in the making, to make access to PubChem easier and conceptually simpler to achieve. For those who would rather learn directly about the inner workings of PubChem data processing and analysis, a C++ API, based on the NCBI C++ toolkit [28], will be made available.

PubChem is a significant source of information on the biological properties of small molecules. The offering of tools and services associated with the access and mining of this data makes PubChem important to the work of scientists worldwide as an enabling resource. PubChem continues to grow and evolve as a function of time. New tools and services are in development and existing offerings are being refined. Feedback from the user community is an important and welcome part of this process to ensure the utility of PubChem to the community is maximized. The NCBI help desk (email: info@ncbi.nlm.nih.gov) is the primary locus for such input.

ACKNOWLEDGMENTS

This research was supported, in part, by the Intramural Research Program of the NIH, National Library of Medicine.

REFERENCES

1. http://pubchem.ncbi.nlm.nih.gov.
2. http://nihroadmap.nih.gov/molecularlibraries/.
3. Benson, D., Karsch-Mizrachi, I., Lipman, D., Ostell, J., Wheeler, D. GenBank. Nucleic Acids Res. 2007, 35, D21–5.
4. Boeckmann, B., Bairoch, A., Apweiler, R., Blatter, M.-C., Estreicher, A., Gasteiger, E., Martin, M.J., Michoud, K., O'Donovan, C., Phan, I., Pilbout, S., Schneider, M. The Swiss-Prot Protein Knowledgebase and its supplement TrEMBL in 2003. Nucleic Acids Res. 2003, 31, 365–70.
5. Berman, H., Westbrook, J., Feng, Z., Gilliland, G., Bhat, T., Weissig, H., Shindyalov, I., Bourne, P. The Protein Data Bank. Nucleic Acids Res. 2000, 28, 235–42.
6. http://www.ncbi.nlm.nih.gov.
7. http://www.ncbi.nlm.nih.gov/sites/gquery.
8. http://pubmed.gov.
9. Chen, J., Anderson, J., DeWeese-Scott, C., Fedorova, N., Geer, L., He, S., Hurwitz, D., Jackson, J., Jacobs, A., Lanczycki, C., Liebert, C., Liu, C., Madej, T., Marchler-Bauer, A., Marchler, G., Mazumder, R., Nikolskaya, A., Rao, B., Panchenko, A., Shoemaker, B., Simonyan, V., Song, J., Thiessen, P., Vasudevan, S., Wang, Y., Yamashita, R., Yin, J., Bryant, S.H. MMDB: Entrez's 3D-structure database. Nucleic Acids Res. 2003, 31, 474–7.
10. OEChem, version 1.5.1, OpenEye Scientific Software, Inc., Santa Fe, NM, USA, http://www.eyesopen.com, 2007.
11. Stein, S., Heller, S., Tchekhovskoi, D. An Open Standard for Chemical Structure Representation: The IUPAC Chemical Identifier. Proceedings of the 2003 International Chemical Information Conference (Nimes), Infonortics 131–43. http://www.iupac.org/inchi/.
12. Lexichem, version 1.6, OpenEye Scientific Software, Inc., Santa Fe, NM, USA, http://www.eyesopen.com, 2007.
13. http://dtp.nci.nih.gov/webdata.html.
14. Richard, A., Williams, C. Distributed Structure-Searchable Toxicity (DSSTox) public database network: A proposal. Mutat Res. 2002, 499, 27–52. http://www.epa.gov/nheerl/dsstox/.
15. Liu, T., Lin, Y., Wen, X., Jorrisen, R., Gilson, M. BindingDB: a web-accessible database of experimentally determined protein-ligand binding affinities. Nucleic Acids Res. 2007, 35(Database Issue), D198–201. http://www.bindingdb.org/.

16. Lipinski, C., Lombardo, F., Dominy, B., Feeney, P. Experimental and computational approaches to estimate solubility and permeability in drug discovery and development settings. Adv. Drug Del. Rev. 2001, 46, 3–26.
17. http://www.daylight.com/dayhtml/doc/theory/theory.smarts.html.
18. Dalby, A., Nourse, J., Hounshell, W., Gushurst, A., Grier, D., Leland, B., Laufer, J. Description of several chemical structure file formats used by computer programs developed at Molecular Design Limited. J. Chem. Inf. Comput. Sci. 1992, 32, 244–55.
19. Tanimoto T. IBM Internal Report 17th Nov. 1957.
20. Durant, J., Leland, B., Henry, D., Nourse, J. Reoptimization of MDL keys for use in drug discovery. J. Chem. Inf. Comput. Sci. 2002, 42, 1273–80.
21. http://eutils.ncbi.nlm.nih.gov/entrez/query/static/eutils_help.html.
22. http://taverna.sourceforge.net/.
23. http://accelrys.com/products/scitegic/.
24. http://www.python.org/.
25. http://www.ruby-lang.org.
26. http://www.perl.org/.
27. http://www.w3.org/TR/wsdl.
28. http://www.ncbi.nlm.nih.gov/IEB/ToolBox/CPP_DOC/.

restrained electrostatic potential, 1, 92, 93
restricted Hartree–Fock (RHF), 1, 46, 48–50
restricted-active-space self-consistent field
    (RASSCF) method, 1, 47
REU *see* Research Experiences for
    Undergraduates
RHF *see* restricted Hartree–Fock
RISM, 2, 266, 267
ROC curve, 2, 297, 306, 307, 315
ROCS, 2, 318
Roothaan–Hall equations, 1, 6–8
rotational–vibrational
    energy levels, 3, 159
    spectra, 3, 169
    transitions, 3, 159
rovibrational eigenvalues, 3, 157
Ru(bpy)$_3^{2+}$, 1, 7
Runge–Gross theorem, 1, 27

S$_N$A, 2, 270, 271
S$_N$Ar, 2, 268–270, 275
sampling barriers, 1, 242, 243
SAR *see* structure–activity relationships
scads, 1, 250
scaling methods, 1, 6–8
Schrödinger equation, 1, 3–15; 2, 297–299, 313,
    314, 316, 318–320
scoring functions, 1, 119–126
scoring functions, quality, 2, 161, 162
self-consistent field (SCF) methods, 1, 6–10,
    37, 46, 47, 53
self-consistent reaction field (SCRF), 1, 118,
    121
self-extracting databases, 1, 223, 225
semantic Wiki, 3, 110, 123, 126–128, 131
semi-empirical methods, 1, 12, 13, 15, 31, 32
    PDDG/PM3, 2, 264, 265, 267, 268, 272, 274,
    276
sextic force fields, 3, 162
SHAKE algorithm, 2, 222
signal trafficking *see* kinome targeting
similar property principle, 2, 141
Slater geminal methods, 2, 28, 30
Smac, 2, 206, 208, 209
small molecule solvation, 3, 50
"soft core" Lennard-Jones interactions, 3, 47
solubility, 1, 135–137
solvation, 1, 117–119, 247
space group symmetry, 3, 94
spectroscopic accuracy, 3, 157
spectroscopic network (SN), 3, 159
spherical harmonics, 3, 167
spin-flip methods, 1, 53
standard domains, 2, 53, 57, 59, 64, 68, 69, 71,
    73–76

standard pK_a, 3, 4
standard uncertainty (su), 3, 87
statistical computational assisted design
    strategy (scads), 1, 250
Steepest Descent Path (SDP), 3, 19
stochastic difference equation in length
    (SDEL), 3, 17–19
    advantages, 3, 20
    disavantages, 3, 20
stochastic difference equation in time (SDET),
    3, 17
Stochastic Gradient Boosting, 2, 137
stochastic models, 1, 215–220
storage capacity, 1, 224, 225
string method, 3, 16
strong pairs, 2, 59, 62, 63, 68, 69, 71, 73, 75, 77
structural mimicry, 3, 217
structural motifs, 3, 211
structure–activity relationships (SAR), 1, 91,
    133–151
structure-based design, 2, 197, 202, 205, 209
structure-based drug design, 1, 114, 120, 125
structure-based hybridization, 1, 191, 192
structure-based lead optimization, 1, 169–183
    application to specific targets, 1, 179
    compound equity, 1, 171
    discovery, 1, 171–175
    fragment positioning, 1, 175–177
    high-throughput screening, 1, 171, 172
    library enumeration, 1, 178
    ligand–target complex evaluation, 1, 178,
    179
    modification, 1, 175–179
    molecular simulation, 1, 177, 178
    structure visualization, 1, 175
    virtual screening, 1, 169, 172–175
structure-based ligand design, 2, 184
structure-based virtual screening, 2, 284
structure-property relationships, 2, 142
substrate access, P450, 2, 178
substrate prediction, P450, 2, 172
support vector machines, 1, 137, 145; 2, 128,
    149
surface diffusion, 3, 138, 140
Surflex, 2, 161
Sutcliffe–Tennyson triatomic rovibrational
    Hamiltonian, 3, 167
symbolic computation engines (SCE), 1,
    221–235
    advanced application-specific procedures,
    1, 229–231
    computation power, 1, 228, 229
    emulation of professional software, 1,
    229–231